The National Trust Book of

British Wild Animals

JONATHAN CAPE
THIRTY BEDFORD SQUARE LONDON

First published 1984

Jonathan Cape Ltd, 30 Bedford Square, London WC1B 3EL
in association with the National Trust and the
National Trust for Scotland.

British Library Cataloguing in Publication Data
The National Trust book of British wild animals.
1. Zoology—Great Britain
I. Burton, John A.
591.941 QL255

ISBN 0-224-02104-4

Printed in Great Britain by
W. S. Cowell Ltd, Butter Market, Ipswich

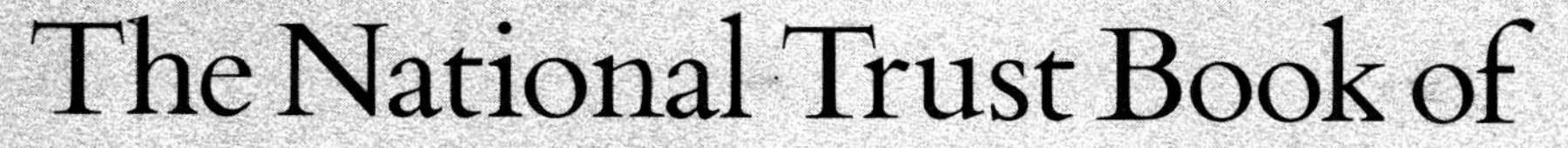

The National Trust Book of

British Wild Animals

edited by John A. Burton
on behalf of the Fauna and Flora Preservation Society

A puffin alighting with a sand eel in its beak.

Contents

Editor's Preface

The various chapters to the *National Trust Book of British Wild Animals* have been contributed by well-known experts. Each has given an over-view of the subject, without attempting to be over-exhaustive or to provide a field guide – such information is given in a list of Further Reading. English names are used throughout, and should the reader want to know the Latin names of any species they can be found in the field guides.

The National Trust was formed only a few years before the Fauna and Flora Preservation Society. In 1903 a group of eminent naturalists, colonial governors and parliamentarians formed the Society for the Preservation of the Wild Fauna of the Empire, which in 1980 became the Fauna and Flora Preservation Society. Although the ffPS has never been a land-owning charity, from its inception it has been actively involved in promoting game reserves, national parks and nature reserves in all parts of the world. It was therefore very appropriate that the ffPS should co-operate with the National Trust to produce this book.

JOHN A. BURTON
Executive Secretary, ffPS

Foreword

Sir Peter Scott

The National Trust for Places of Historic Interest or Natural Beauty and the Fauna and Flora Preservation Society were both founded within a decade of each other at the turn of the century – the former in response to the need to preserve Britain's natural and historic heritage, and the latter to preserve the wildlife of the Empire. The National Trust for Scotland was founded in 1931. One of the first acquisitions of the National Trust was Wicken Fen in 1899, one of Britain's finest nature reserves. Although most people probably associate the Trusts with historic buildings and stately homes, they are in fact two of the largest holders in the UK of nature reserves and other outstanding areas for wildlife, ranging in size from tiny reserves only a few acres in extent up to the thousands of acres of the Scottish estates. To a naturalist, it is particularly pleasing to see the recent emphasis which the National Trust has placed on its holdings of countryside and areas of natural beauty. It is reassuring that over 500,000 acres owned and a further 75,000 acres protected by covenants are safe from the more undesirable developments taking place in the countryside. The contribution the National Trust makes is impressive: of its 500,000 acres over 200,000 have been listed by the Nature Conservancy Council as Sites of Special Scientific Interest; these SSSIs have been designated to include a representative selection of Britain's most important wildlife habitats. It is by acquiring such land that nature can really be held in trust for future generations.

People often ask why there are so many different conservation charities and if they are all really necessary. This book, it seems to me, helps to answer this question. While each organisation has its own specialised role, a spirit of co-operation exists between many of them. In this book, the expertise of an international wildlife society has been able to assemble an informative text, which I hope will stimulate the visitor to Trust properties to take a fresh look at their wildlife.

Guillemots and fulmars on The Stack, Inner Farnes.

I have a strong personal interest in some of the problems confronting the National Trust. For example, the conflict between providing public access and the need to protect wildlife from disturbance is a problem which we have also encountered at the Wildfowl Trust, at Slimbridge and elsewhere. The National Trust's solution to this conflict, which was first developed by John Bailey, Chairman in the 1920s, is to make the preservation of wildlife the top priority. This may mean that you will find parts of a reserve out of bounds for part of or even the whole of the year, but we should all respect the need, sometimes, for total protection of wildlife; for me it is enough to know that it is there.

As the world's tropical forests disappear to become chipboard or to make way for the grazing of beef cattle, one can only regret that the National Trust does not have an international counterpart, although there are equivalent organisations in the United States, Australia and elsewhere, and in many parts of the world the National Parks serve the same purpose. At this time the importance of land acquisition, and preservation, and the power to declare it 'inalienable' is vitally important if we are to pass on to our children the still significant richness of wildlife we inherited from our parents.

March 1984 PETER SCOTT

Acknowledgments

The authors and publishers are grateful to the following for their kind permission to reproduce colour plates on the pages as listed: J. A. Burton, p. 62 left, p. 78 below left, p. 90 above right and below left; Robert Burton, p. 9, p. 27; Michael Clark, p. 23; Bruce Coleman Ltd, p. 47 (photo by Hans Reinhard), p. 103 (photo by Dennis Green), p. 107 above (photo by Andy Purcell) and below (photo by Dennis Green), p. 110 left (photo by Gordon Langsbury), p. 111 (photo by Roger Wilmshurst), p. 118 right (photo by Jane Burton), p. 130 (photo by Jane Burton), p. 131 (photo by Ronald Thompson/Frank W. Lane), p. 146 right (photo by Jane Burton); D. Element, p. 12, p. 16 right, p. 22 left, p. 30 left, p. 34 left and right, p. 63 left and right, p. 74 left and right, p. 75, p. 78 above left and right, below right, p. 115, p. 119, p. 122 above, p. 138, p. 139, p. 142, p. 143 above left, below left and right, p. 150 below, p. 151 above and below; E. H. Herbert, p. 59, p. 66, p. 67, p. 71, p. 79, p. 87, p. 90 above left, below right, p. 94, p. 143 above right, p. 146 left, p. 150 above, p. 154; G. Kinns, p. 31; A. R. Martin, p. iv, p. 50, p. 51; P. Morris, p. 22 right, p. 30 right, p. 38, p. 43, p. 62 right, p. 122 below, p. 147, p. 155, p. 158 left and right, p. 159, p. 162, p. 163 left and right; the National Trust, p. viii (photo by Mike Williams), p. xii (photo by Mike Williams), p. 5 left (photo by Mike Williams) and right (photo by Alan North), p. 13 (photo by J. M. Hannavy), p. 35 (photo by Mr Bisserot), p. 39 (photo by R. E. Stebbings), p. 86 (photo by Mike Williams), p. 99 (photo by Mike Williams), p. 127 (photo by R. Hillgrove), p. 166 (photo by Mike Williams); J. Reed, pp. 54–5, p. 95, p. 98 left and right; R. E. Stebbings, p. 46; C. S. Waller, p. 16 left, p. 110 right, p. 114.

Black and white photographs are reproduced by kind permission of the following: J. A. Burton, p. 15 right, p. 18 above and below, p. 21 left, p. 26 right, p. 65, p. 76 left and right, p. 141, p. 156, p. 157; Michael Clark, p. 11, p. 14 left and right, p. 21 right, p. 24, p. 28, p. 36, p. 40, p. 45 left, p. 120; Bruce Coleman Ltd, p. 37 (photo by Jane Burton), p. 85 (photo by Kim Taylor), p. 105 (photo by P. A. Hinchliffe), p. 108 (photo by L. R. Dawson), p. 128 (photo by Jane Burton), p. 135 (photo by Jane Burton); D. Element, p. 61 above and below, p. 64, p. 70 right, p. 72 left, p. 80, p. 97, p. 121 left and right, p. 144, p. 148, p. 149; A. R. Martin, p. 49, p. 52, p. 57, p. 101 left; P. Morris, p. 26 left, p. 70 left, p. 93, p. 109, p. 117, p. 137 left and right, p. 161, p. 165 left and right; the National Trust, p. 2 (photo by Mike Williams), p. 15 left, p. 45 right (photo by R. E. Stebbings), p. 82 left and right; J. Reed, p. 72 right, p. 92, p. 101 right; C. S. Waller, p. 104.

Marshes at Salthouse, north Norfolk – one of several important East Anglian reserves owned by the Trust.

1 *The National Trust and Wildlife*

Dr T. W. WRIGHT

Although many people understandably still associate the National Trust primarily with the preservation of country houses and gardens, it is also now a very large private landowner, protecting over half a million acres of farmland and open country in trust for the nation. Through its ownership of such a wide diversity of habitats – woods, lakes, lowland pastures and upland moors, sea cliffs, sand dunes and marsh – the Trust has become one of the major organisations concerned with nature conservation in Britain. An observer with a keen eye and plenty of time for travel should be able to find all the animals mentioned in this book somewhere on National Trust property in England, Wales or Northern Ireland, or on similar properties belonging to the National Trust for Scotland.

The National Trust was founded in 1895 by three private individuals who foresaw that the spread of industry and the growth of towns would be an increasing threat to the countryside and its historic buildings. Miss Octavia Hill was already nationally known for her pioneer achievements in the field of housing reform; Sir Robert Hunter was a professional administrator, Solicitor to the Commons Preservation Society and to the General Post Office, whose special concern was for the open spaces of Surrey; the Reverend Hardwicke Rawnsley was Canon of Carlisle, a man of wide interests with a passionate devotion to the incomparable scenery of the Lake District. It was their combined genius to see the need for a national organisation to preserve, through ownership, places of historic interest or natural beauty for the benefit and enjoyment of all.

The first National Trust Act of 1907 made the National Trust a statutory body with the unique power to declare its property 'inalienable'. This principle of inalienability has been fundamental to the Trust's development ever since. Briefly, it means that such property cannot be sold or given away, or even compulsorily acquired by government departments or any other authority, without special resort to Parliament. It gives the greatest possible guarantee that land once vested in the Trust and declared inalienable will indeed be held by the Trust in perpetuity on behalf of the nation.

The 1907 Act includes as one of its objectives of landownership 'the preservation (so far as practicable) of their natural aspect features and animal and plant life'. This must place it among the earliest enactments in any country specifically establishing an agency to acquire and hold land for the purposes of nature conservation, and it is clear evidence of the concern of the Trust's founders with this aspect of their work. They undoubtedly appreciated that the phenomena of nature were part of the wider concept of beauty which they wished to preserve, and some early acquisitions were made in which nature conservation was the main objective. The first of these was Wicken Fen (1899), already recognised for the variety of its plant, bird and insect life and now virtually all that remains undrained of the once extensive fens of the Cambridgeshire Great Level. This was followed by Blakeney Point (1912), Scolt Head (1923), and the Farne Islands (1925). In all three, their value as sea-bird nesting sites was at least as important as their remote beauty. Nor was the initiative always taken by the Trust; it was often approached by local groups or individuals who had long studied and cherished the wildlife of a particular area. The Ruskin Reserve, four and a half acres of marshy woodland in Oxfordshire, was given in 1917 by the Ashmolean Natural History Society, and is typical of several similar properties acquired in this way.

A distant view of the Farne Islands from St Aidan.

But in general in those early days it was not wildlife but natural beauty, and especially beauty threatened by urban development, which was the most common criterion applied to the acquisition of open land. In the 1920s and early 1930s particularly, much was heard of the growing problems of 'ribbon development' and the unchecked urban sprawl, but they were perceived mainly in terms of landscape and aesthetic values rather than as direct threats to wildlife. As a national issue, nature conservation did not gain prominence until the end of the Second World War and the publication of the White Paper *Conservation of Nature in England and Wales* in 1947. This led directly to the establishment of the Nature Conservancy in 1949, as a new government agency responsible for all aspects of wildlife policy. The Conservancy soon expressed an interest in Trust properties. Fourteen of the seventy-three proposed Nature Reserves listed in the White Paper belonged to the Trust, and Scolt Head and Cwm Idwal were declared National Nature Reserves in 1954, followed by Worms Head in 1957. Today the Trust owns thirteen National Nature Reserves, eleven leased to the Nature Conservancy Council (which succeeded the Nature Conservancy in 1973) and two managed by the Trust itself under a Nature Reserve Agreement.

Until the mid-1970s the Trust's general response to increased public interest in nature conservation was to make arrangements for other bodies with more specialised knowledge to manage those of its properties which were known to be important wildlife habitats. This was logical, since at that time the Trust's field staff was extremely small in relation to the area involved, and was wholly occupied in the work of general estate management. In addition to the arrangements with the Nature Conservancy, leases and management agreements covering some forty properties were concluded with the County Naturalists' Trusts, which expanded rapidly from the early 1960s to cover virtually the whole country within the decade. A close and effective working partnership was established, combining the advantages of National Trust ownership with the specialised local knowledge and enthusiasm of the County Trusts, and this partnership continues to prosper.

The publication of the Nature Conservancy Council's *Nature Conservation Review* in 1977 marked a significant turning point in the Trust's attitude to nature conservation. This extremely important work, which discusses in great detail the whole concept of a national approach to nature conservation and lists all the so-called Sites of Special Scientific Interest declared by the Council up to that time, brought into sharp relief the immense biological value of the Trust's holdings. No less than 342 Trust properties, covering some 203,000 acres, or almost half the Trust's total land holding, are included in these lists. The Trust was thus shown to be easily the largest private owner of land considered by the Council to be of nature conservation importance. This is all the more surprising when one considers that the great majority of these properties had been acquired by the Trust primarily for their intrinsic beauty rather than as wildlife habitats, although it clearly demonstrates the close parallels between these two.

The Trust clearly could not contemplate delegating the management of anything like this proportion of its land to other organisations; this would amount to an abdication of its responsibilities totally incompatible with its objectives of ownership. While continuing its policy of joint responsibility with the Nature Conservancy Council, the County Trusts, the Royal Society for the Protection of Birds, and other specialist bodies through leases and management agreements for the more important sites, the Trust has therefore resolved that nature conservation must also play a greater part in its own general property management. It is now apparent that the relatively low-key, preservationist approach to land management which the Trust has always practised in pursuit of its primary objective of landscape conservation is in many cases also likely to favour the maintenance of wildlife habitats. But the pressure on such areas, already acute and likely to become still more severe in the future, now requires a more positive attitude. The conscious inclusion of wildlife conservation into the Trust's management planning has therefore spread from the more specialised sites to become an important element in the conservation of all its properties.

In addition to the preservation of the landscape and wildlife conservation, the Trust's management objectives also include the provision of public access and the maintenance of its farms and woodlands as viable economic units, generating income both for its tenants and for the Trust itself. These four major objectives are clearly to some extent incompatible, and the whole essence of the Trust's management policy must be to arrive at the correct balance between them in each individual case.

While some of the factors involved are difficult or even impossible to quantify, it is possible to arrive at a satisfactorily objective assessment of biological importance by means of adequately detailed surveys. Since 1979 the Trust has employed its own biological survey team to provide its management staff with this information, gathered both from its own field work and from existing data collated from a wide variety of specialist sources. When complete, these surveys will cover not only the nature reserves and other sites of known scientific importance but virtually all the Trust's land, and they will be added to and updated as new properties are acquired. They are not management plans in themselves; their purpose is to provide the essential biological information on which the plans can be based, and to suggest how nature conservation can best be integrated into the wider management objectives.

There are certain broad principles which guide the Trust in deciding priorities. The concept of natural beauty which is so fundamental to the Trust's purpose requires that in almost all cases priority is given to landscape preservation. In woods and open country, management is directed towards maintaining the status quo; to reducing change to the inescapable minimum implicit in the gradual growth and decay of all living systems. On enclosed farmland, which the Trust does not keep in hand but leases to tenants under the terms of an agricultural tenancy, preservation can best be achieved by maintaining as far as possible the traditional farming practices which have moulded the landscape, and in particular the hedgerow patterns and the scattering of trees and small copses which are its principal components. This inherent characteristic of resistance to change is itself a powerful factor operating in favour of wildlife conservation, particularly in the case of

Left *Golden Cap Estate, Dorset.*

Right *The Giant's Causeway viewed from above, looking west along the North Antrim cliff path.*

animal communities. In general, the first requirement of the larger animals and birds is freedom from disturbance, especially in the breeding season. Given this, they can adapt to quite wide variations in their physical environment. Thus otters, whose numbers have decreased alarmingly in recent years owing in great part to the effects of disturbance, find sanctuary on Trust properties where their favoured riverbanks are designated as 'otter havens' and specially protected. And by preserving old buildings – not only great houses but also innumerable examples of rural vernacular architecture – the Trust acts as host to numerous colonies of bats, which elsewhere have been dispersed and seriously depleted by the destruction of their roosts when buildings have been demolished.

While some animals need help to survive under modern conditions, others are sometimes too successful and their populations must be controlled to keep them at an acceptable level. Roe deer are particularly fond of dense young forestry plantations, where they damage young trees and raid the surrounding farm crops. The small muntjac deer are now spreading rapidly in lowland Britain, and play havoc with gardens round the Trust's Ashridge estate in Hertfordshire. But wherever control is necessary, the Trust seeks to maintain as large a population as possible without the level of damage becoming intolerable. In fact, surprisingly large numbers can be accommodated if the correct control measures are applied.

Smaller animals are much more dependent on specific habitats which are often fragile, localised and particularly at risk from changing factors in the environment. The butterflies of the chalk downlands are an outstanding example; the short sward containing their food plants was formerly maintained entirely by grazing, by sheep and rabbits. Both have declined in recent years, the sheep in response to changes in farming practice and the rabbits victims of myxomatosis, and as a result many thousands of acres have been invaded by coarser grasses and thorn scrub. The Trust is engaged in extensive programmes of scrub clearance and mowing, and, increasingly, has re-introduced sheep grazing, with careful control of numbers and grazing periods. Certain reptiles and amphibia, too, need special conditions which are becoming increasingly scarce, and on appropriate sites the Trust is carrying out quite detailed and specific programmes of habitat manipulation for the benefit of such species as sand lizard, smooth snake and natterjack toad.

To cope with the extra work entailed in this more active approach to conservation management, the Trust has increased its field staff by the appointment of many new property wardens. Acting on the instructions of the land agents in each of the Trust's sixteen regions, they carry out all the practical work specified in the property management plan. In addition, and increasingly on the more important conservation sites, they are becoming involved in the monitoring and recording of populations and changes in their habitats in response to these activities. The Trust is not a research organisation, but these monitoring programmes are of the greatest importance on those properties on which change is taking place, whether deliberately induced or due to other causes, if a sound basis for future management is to be established.

Practical habitat management is labour intensive and time-consuming, and the wardens alone often cannot accomplish unaided all that is required. The Trust is specially fortunate in the voluntary help it receives from all quarters, and particularly from young people through its Acorn Camp schemes and Young National Trust groups. Acorn Campers stay for a week or longer on a property, and carry out a wide range of conservation work under the direction of the local staff. YNT groups are beginning to develop in several Trust regions. They perform a similar function to the Acorn Groups but on a more permanent basis, often 'adopting' a property within their area and treating it as their special responsibility, or tackling more ambitious projects which need a more sustained effort. Both are extremely useful, and we like to think that the volunteers who give their time and efforts so freely to these worthwhile tasks derive satisfaction from seeing the improvements which they are able to achieve.

It is also fundamental to the Trust's philosophy that once their preservation has been assured its properties should be enjoyed as fully as possible by its members and by the general public. This has been implicit in the concept of Trust ownership since its earliest years, when the founders regarded its function in this respect as the provision of 'lungs' where the urban public could refresh themselves in the beauty and tranquillity of unspoilt and protected countryside. There was little reason to suppose in those early days that the sheer number of visitors would pose a threat to the very things they had come to enjoy, yet this is now the case on many properties in popular holiday areas or near centres of population. Management today must concern itself closely with measures to mitigate the effects of visitor pressure, both to protect the landscape from physical erosion and, more particularly, to safeguard wildlife habitats. In wetlands and sand dunes for example, where the ground is specially vulnerable to trampling, board walks and similarly strengthened routes must be provided, and paths must be kept well away from nesting sites and other fragile areas. Herons are extremely sensitive in this respect, and woodland paths near heronries often have to be closed when the birds return to them in early spring, though they may be freely used once the young are fledged. Elsewhere, with careful planning, surprisingly large numbers of people can be accommodated without danger. Boatloads of visitors come to Blakeney Point every year to view the tern colonies from the hides set into the shingle bank, and many thousands are able to see at close quarters the dense colonies of nesting sea-birds on the Farne Islands because access is controlled and limited to certain times of the day. On all but the most sensitive sites nature conservation is fully compatible with quite a high degree of public access provided the possibility of damage is appreciated and precautions are taken. Often it is necessary only to provide and maintain a good footpath network, but much more can be achieved if contact can be made with the visitor so that the reasons for such restrictions as have to be imposed can be explained.

As the pressure for access increases, this need to make contact with and inform the visitor becomes correspondingly more important. The great majority of us now live in an urban environment with few opportunities for direct experience of the natural world, and need help to appreciate it fully. We hope that this book and similar publications will encourage you to visit Trust properties and enable you to appreciate them more fully through a deeper understanding of their natural history. Once there, your visit may be made more enjoyable by the provision of a simple interpretative leaflet or even a nature trail, but such things can easily be overdone and become an annoying intrusion. One of the delights of a visit to Trust property is that you are often not confined to a set route, but can wander at will down any one of a number of different paths and discover for yourself far more than a leaflet would have space to describe. None the less, trails and leaflets can be of value, particularly to school parties, who are increasingly turning to Trust properties to illustrate the theoretical principles they have studied in the classroom. A wide range of leaflets are on sale in shops and information centres in all the Trust's regions, and on some properties there are other facilities. At Witley Common in Surrey, for example, the information centre opened in 1976 has been visited by many hundreds of school parties, who learn from the static displays and an audio-visual programme the sequence of events which has moulded the Common as they see it today.

The National Trust is a charity, relying heavily on the generosity of its donors and the subscriptions of its members, who now number more than a million. But a substantial part of the income needed to further its work of conservation is derived from the rents which it charges the tenants of its farmland. It is here that the Trust's special objectives need to be most clearly defined and as far as possible quantified in economic terms, so that they can be taken into account when the rents are agreed. The Trust could not continue to preserve its very extensive agricultural landscapes without the co-operation and understanding it receives from its farm tenants. Over many years, the Trust has demonstrated time and again that wildlife conservation, public access and the preservation of beautiful landscape can all be integrated into a profitable farm economy. But there must be good will on both sides and a clear recognition that these special objectives, which in the Trust's case are of paramount importance, entail a level of return which is less than the maximum possible. It is fashionable nowadays to present the options for countryside management in terms of a stark, black-and-white contrast between two diametrically opposed entrenched positions: excessively restrictive nature conservation on the one hand and unbridled exploitation using all the resources of modern technology on the other. This may make colourful headlines, but in the great majority of cases a compromise is possible which does not represent the lowest common denominator of both attitudes but a genuine way forward based on a balanced appraisal of all the factors involved. The Trust's tenants accept the limitations of the Trust's objectives, and in turn the Trust endeavours to provide, subject to these constraints, the conditions within which they can maintain a viable farming enterprise.

The place of nature conservation in Britain is still a matter for lively debate, but recent legislation and the growing authority of the Nature Conservancy Council and the Countryside Commission are signs of its wider acceptance in the future. The Trust freely acknowledges the assistance it has received from these two government agencies, and hopes that as conservation is increasingly accepted as an integral part of land management its properties may point the way for others to follow. All the wide diversity of wildlife described in this book has survived over centuries, by a process of adaptation to gradual changes in farming practices and land management. In recent years the rate of change has enormously increased, and it is largely the rate of change which governs the survival of wildlife communities. The Trust believes that change can be contained by consistently adopting a balanced approach in which landscape preservation and nature conservation play a full part. It gladly accepts the duty imposed by its founders to preserve all that is best of the past and hand it on to succeeding generations.

A common seal pup.

PART ONE

Mammals

2 *Deer, Cattle, Goats, Sheep and Ponies*

JOHN A. BURTON

One of the finest collections of deer is to be found at Knole, which is owned by the National Trust, near Sevenoaks in Kent, some twenty-five miles from London. Knole is set in a park, only partly owned by the Trust, with clumps of mature oaks, beeches and coverts of bracken in the low hills and valleys which provide an ideal setting for deer. Fallow and sika deer are to be seen there. Petworth in Sussex is another National Trust property known almost as much for its deer as for its stately home.

Throughout Britain the occurrence of deer can be roughly divided into park deer and truly wild and feral (descended from captive) deer. But these two distributions are closely linked, since many of the wild populations started as escapes from parks. The National Trust owns seven deer parks but thousands of acres of land harbour wild deer.

There are two species of deer native to Britain: the red and roe deer, and four widely distributed introduced species: the fallow, muntjac, Chinese water deer and sika deer. Others such as Père David's deer, axis deer and wapiti are quite numerous in a few deer parks such as Woburn and Whipsnade and there is a herd of reindeer in the Cairngorms.

RED DEER

The largest of Britain's deer is the red deer, which although primarily a woodland species is also the 'stag' of the Scottish Highlands. The red deer of Britain are considerably smaller than those living in the wooded mountains of eastern Europe, and in New Zealand (where they have been introduced). In Britain a stag over 150kg (360lb) is exceptionally large, whereas in eastern Europe they reach over 255kg (560lb).

The deer park at Charlecote in Warwickshire, which was owned by the Lucy family from the thirteenth century until it passed to the National Trust in 1945, is the home of a herd of red deer, which are the descendants of those which William Shakespeare is supposed to have poached; legend has it that he was fined by Sir Thomas Lucy in 1583.

Outside deer parks the red deer is most widespread in Scotland, including many of the islands, and in north-west England. Elsewhere in England, Wales and Ireland most of the red deer to be found have escaped from deer parks and subsequently bred in the wild or have escaped from hunts. Until well into the present century it was a common practice in areas without deer for them to be carted and hunted with staghounds and when brought to bay to be recaptured and kept for another hunt. Sometimes these deer escaped and formed the nucleus of a breeding population. The red deer living in Dunwich Forest in eastern Suffolk and the adjacent National Trust property of Dunwich Heath may have originated in this way.

FALLOW DEER

The fallow deer is the most popular of the park deer. Although originally a Mediterranean species, the fallow deer has been present in Britain since the Middle Ages and some of the original introductions, such as those in Cannock Chase, Hatfield Forest, Epping Forest and the New Forest are still living in a semi-wild state. The fallow deer occurs in a number of colour varieties, which range from almost pure white to black, but the normal summer coat is a rufous-fawn with white dappling on the back and sides. Like most other deer,

During the breeding season or 'rut' the red deer male makes a belching call known as 'belling' and defends a harem of females.

Magpies and other birds are often attracted to deer, feeding on insects which have been disturbed by them while they graze.

Red deer stags in Lyme Park, Cheshire. The antlers are shed each year, and when they regrow are covered with a soft 'velvet' which later dries and is scraped off.

only the males carry antlers, which in the case of the fallow deer are flattened – palmate – unlike the widely branched antlers of the red and sika deer, or the short spikes of the roe and muntjac.

In some ways, the best time to see park deer is in the autumn when the males are in rut, grunting and bellowing at each other, and charging and clashing horns in territorial disputes. However, caution should always be exercised and they should be viewed from a discreet distance as they are potentially dangerous to any intruders, including humans.

SIKA DEER

The sika deer is slightly smaller than the fallow deer, and also dappled on the back and sides. It originated in the Far East both on the continent of Asia and the islands of Japan, Taiwan and elsewhere. In some areas feral sika deer have hybridised with red deer. Although not particularly widespread outside deer parks, sikas are now found in scattered populations from Dorset to Inverness, in Ireland, and on the National Trust owned Island of Lundy, and Brownsea Island in Poole Harbour, Dorset. A large herd can also be seen alongside fallow deer in the grounds of Knole; smaller herds occur in the grounds of a number of other stately homes.

ROE DEER

The native roe deer is probably the most widespread species in England and Scotland, but is not found in Wales and Ireland. It occurs in three main areas: southern England from eastern Cornwall to mid-Sussex, East Anglia (as a result of introductions) and northern England and Scotland. It is a small, shy species which, unlike the sika, red and fallow, does not live in large herds, and is therefore unsuitable for keeping in deer parks. It is, however, often locally abundant in woodlands and probably occurs on many – if not most – of the larger National Trust properties within its range.

OTHER DEER

The muntjac and Chinese water deer are very small – not much bigger than a cocker spaniel – and the existence of both is the result of deliberate introductions or escapes, mainly from the Duke of Bedford's extensive collection of deer at Woburn, dating from around 1900. The muntjac, often known as the barking deer because of its nocturnal,

The characteristic spiky antlers and furrowed brow of the muntjac deer (left) *distinguish it from the Chinese water deer, which lacks antlers but has short tusks.*

Left *Brownsea Island is the home of sika deer and also of one of the last remaining populations of red squirrels in England.*

Right *Sika deer mating in Knole Park, Kent.*

dog-like call, is more often heard than seen and is now quite widely spread over central, southern and eastern England, but the Chinese water deer has a much more restricted range in central and eastern England. The muntjac has only very short spiky antlers and a frowning expression. The Chinese water deer has no antlers, but the males have long tusk-like canines.

Although not strictly wild, reindeer live in the Cairngorms, Scotland. They were introduced in 1952, in the belief that reindeer had become extinct in Britain round about the twelfth century. In fact, although reindeer had lived in Britain during and after the Ice Ages, it is doubtful if they survived as recently as the twelfth century. Bones found in medieval sites in Scotland are probably derived from venison imported by Vikings from Scandinavia.

Much of the terminology to do with deer is connected with the elaborate ritual associated with medieval hunting. The language used to describe deer is often derived from the Norman French used by the medieval nobility (for example, venison is the term for deer meat). In much the same way the appellations of domestic animals are Anglo-Saxon in origin when they are alive (and being iooked after by the serfs) but are Norman French once they are cooked and eaten: ox, but beef (boeuf); sheep but mutton (mouton); calf but veal (veau).

Deer feed on a wide variety of herbs and grasses and also browse on tree leaves and in hard weather eat shoots and bark. In agricultural areas, therefore, deer may cause some damage, and in young forestry plantations the damage they can do to unprotected trees can be quite extensive. However, in recent years deer have been viewed increasingly not only as an amenity but as an economic benefit. Deer are culled for venison, both in parks and in the wild. In some areas the numbers of deer have risen quite dramatically in recent years, and so has the number of injuries due to deer encountering poachers, snares, barbed

Left *A roe deer kid is hidden by its mother in dense foliage for the first few days of its life until it is strong enough to follow her.*

Right *A fallow buck showing the characteristic palm-shaped (palmate) antlers.*

wire, cars and the many other hazards of the countryside.

Deer parks were originally established to enable their owners to hunt deer more easily. In some of the earliest enclosed parks, such as that of Henry I at Woodstock in Oxfordshire, hunts involving many other exotic animals, as well as deer, were organised. But for the last hundred years or so most deer parks have been maintained mainly for ornamental purposes, with the surplus animals being culled each year to prevent over-grazing.

CATTLE

Truly wild cattle no longer exist in Britain, but for centuries herds of white cattle have been kept on large estates. It was probably during the thirteenth century that such herds were enclosed at Chartley near Uttoxeter and Chillingham in Northumberland.

In appearance the white cattle are probably fairly similar to the extinct aurochs. Their precise origin is uncertain, but by the time the Romans left Britain wild cattle were known in several parts of England and Scotland. References to wild cattle are made in laws promulgated by King Canute (Knut), and although they were undoubtedly protected by the harsh forest laws of William the Conquerer, in 1225 Henry III relaxed the protection and as a result cattle, deer and other game rapidly declined.

These cattle were still very wary and unapproachable even a hundred years ago. It required skill to stalk them, and to hunt them was considered a sporting privilege. In 1872 a bull was shot by the Prince of Wales (later Edward VII) while he was a guest at Chillingham.

The herd at Chillingham is considered to be the purest, but others derive from Cadzow (Lanarkshire), Dynevor (Dyfed), Vaynol (Gwynedd) and Chartley. Animals from several of these herds have now been moved to other locations and may be seen on some National Trust properties in the future. Although these wild cattle are predominantly white (the white colour is a prominent gene) other colours do arise, particularly black.

Highland cattle superficially may appear to be wild cattle, but they are in fact thoroughly domesticated and usually quite docile. However, by crossing various cattle, including types such as Highlands which retain certain primitive characteristics, it has been possible to re-create animals which look rather like the ancestral aurochs.

GOATS AND SHEEP

On North Ronaldsay there are sheep which feed for much of the year on seaweed, but although they live in a remote situation, they are not in any sense wild sheep. They are, however, of considerable interest because of their adaptation to a particular environment. Many breeds of sheep have become rare, as have other domestic animals, owing to modern intensive animal husbandry techniques and an increasing number of these rare breeds are being kept on National Trust properties – they form as much a part of Britain's heritage as buildings and the countryside.

In the more remote parts of northern England, Scotland, Wales and Ireland there are populations of wild goats – some of which are impressive, long-horned, shaggy-coated animals.

PONIES

There are no longer any truly wild horses left in Britain. There are horses or ponies living in a semi-wild state, but these are all descendants of domestic horses which have gone feral. Populations are to be found in the New Forest, Dartmoor, Exmoor, the Lake District, Wales, the Hebrides and elsewhere. Some, such as the New Forest ponies, have been 'improved' from time to time by cross-breeding with Arab stock; the Exmoor ponies are often considered among the most primitive. The 'wild' ponies are generally stocky, sturdy beasts, often with shaggy coats in order to withstand the bleak winters they encounter on the moors and heaths where they live.

Visitors to National Trust parks and other reserves and open spaces where deer, wild ponies and other wildlife survive, should always keep their distance; however tame the animals may appear to be they still retain some wild characteristics, and feeding such animals will often have the added risk of exposing them to fast-moving cars, when they start coming to roads to seek out humans to feed them.

Although not truly wild, ponies live on several of the moors and forests of Britain. This mare and foal are grazing in the New Forest.

Soay sheep from the island of St Kilda in the Outer Hebrides are one of the most primitive of all sheep and thought to be descended from animals imported by Norsemen. They are now seen in many parks, including National Trust properties.

3 *Carnivores*

ROBERT BURTON

Britain, and England and Wales in particular, has an impoverished collection of flesh-eating mammals. During the last Ice Age, the British Isles were the home of the wolverine, Arctic fox and polar bear. As the glaciers retreated and Britain became an island, these animals died out and the newly formed sea prevented the arrival of European mink, stone marten and lynx from continental Europe. Of the carnivores which did arrive, men have since wiped out the brown bear, about a thousand years ago, and the wolf two hundred years ago, while others were persecuted until they became extinct throughout much of the countryside.

Our remaining carnivores are shy, elusive animals. Although they shun contact with humans, they can sometimes be watched undisturbed, either by chance encounter or by carefully lying in wait. Even a fleeting glimpse of one of these animals is rewarding, but often more can be learned by studying tracks, trails and other signs of their activity. These include their footprints, which can be seen in mud, tufts of fur caught on barbed wire, and the remains of their prey. The advantage of such indirect observation is that, especially for our rarest carnivores, the animals are left undisturbed.

RED FOX

The fox has adapted to the taming of the countryside and urban spread better than other carnivores. Traditionally it has either been persecuted for its depredations on poultry, sheep and game, or encouraged so that it can be hunted, but recently it has colonised towns and cities. Lore on foxes has accumulated over the centuries but the real truth about the private life of the fox has been revealed only recently by extensive research with radio tracking and other technical aids.

At one time the fox was held to be one of the few mammals which was monogamous. The dog fox was seen to remain with the vixen while she was rearing the cubs and even to bring food to them. On the other hand, foxes also had a reputation for being solitary nomads. Research has revealed a very variable social system which makes generalisation impossible. Some foxes are 'itinerants' which wander over long distances, but most live in territories, the size of which varies from 20 or so hectares (50 acres) in the wooded countryside of southern England to well over 1,000 hectares (2,500 acres) on the hills of northern England, Scotland and Wales. The owners of a territory range from a single pair of foxes to a group of half-a-dozen adults and their offspring. The group consists of usually one adult male and several vixens. The vixens are probably related, perhaps a mother and her grown-up daughters.

The cubs are born in spring, usually in an underground burrow or 'earth'. It is usual for only the dominant vixen to give birth and the others to act as 'nannies', guarding the cubs and bringing them food. For the first three weeks of life the cubs keep warm by huddling, and the vixen lies with them. Thereafter she rests in a separate place. The first steps outside are taken at six weeks, two weeks after the cubs have started to take solid food. The vixen takes her cubs on hunting trips and they learn to catch prey through practice.

The differences in the social life of the fox are largely a result of variations in abundance and type of food. Foxes are solitary and opportunistic hunters whose diet ranges from deer calves and pheasants to caterpillars and blackberries. Rabbits were the main food before the advent of myxomatosis, but foxes then turned to smaller animals, preferring field voles to other small rodents, and making greater inroads on game-birds. Lamb-stealing is unimportant on a national scale, but it can cause severe losses for the individual farmer and is often the work of a single fox which has developed a taste for lambs.

Traditional trapping and hunting of foxes has had little or no effect on their numbers. Studies have shown that hunts take no more than a regular cull which removes the weaker individuals. The problem of fox-hunting lies in the realm of morality and ethics, and the National Trust often receives instructions from donors of property either to allow or to ban hunting. Potentially a more serious threat to foxes comes from the demand from the fur trade for pelts. Fox fur has become fashionable and the steep rise in prices could stimulate a demand that exceeds the birth rate of foxes and leads to a decline in the population.

STOATS AND WEASELS

The two smallest British carnivores are most often seen when they are crossing the road or peering inquisitively from a stone wall. The stoat is the larger of the two, but females are smaller than males in both species and a female stoat is similar in size to a male weasel. The best distinction is the black-tipped tail of the stoat, which is retained when the animal turns white, as it does regularly in northern Britain in the winter.

Left *Gamekeepers' gibbets – where the stoats, weasels and other trapped predators were hung up – were once a familiar sight in the countryside. However, a more enlightened attitude makes this an increasingly rare sight.*

Right *The weasel is smaller than the stoat and has no black tip to its tail.*

Stoats and weasels are solitary, although it is not unusual to see a female on the move with her youngsters. They are hunters and track down their prey mainly by scent and kill it with a neat bite to the back of the neck. They both live in a wide variety of habitats but typically their hunting grounds are hedges, walls and rough herbage, where there is plenty of cover and an abundance of small animals to capture. Weasels are small enough to enter the runs of mice and voles, and a scarcity of rodents causes a drop in weasel numbers. Male stoats, especially, are too large to enter the runs and they hunt more in open country or invade rabbit burrows.

Occasionally a weasel may kill a rabbit many times its own weight, but stoats regularly prey on rabbits, which were the principal item of their diet until myxomatosis. The stoat throws its victim over and rakes it with its claws while maintaining the neck bite. A hinge prevents the jaw from being dislocated while it is closed on the victim. Both species are good climbers and raid birds' nests in trees. Broods of tits are particularly vulnerable when they are hungry and noisily calling for food.

AMERICAN MINK

Mink have regularly escaped from ranches where they were bred for their fur and, since they were first reported as breeding in the wild, in Devon in 1956, they have spread to many parts of the country. They are still absent from much of Wales, the Midlands and the Lake District. The presence of mink is given away by finding its five-toed footprint near water (smaller than an otter print), and droppings with fishbones that smell

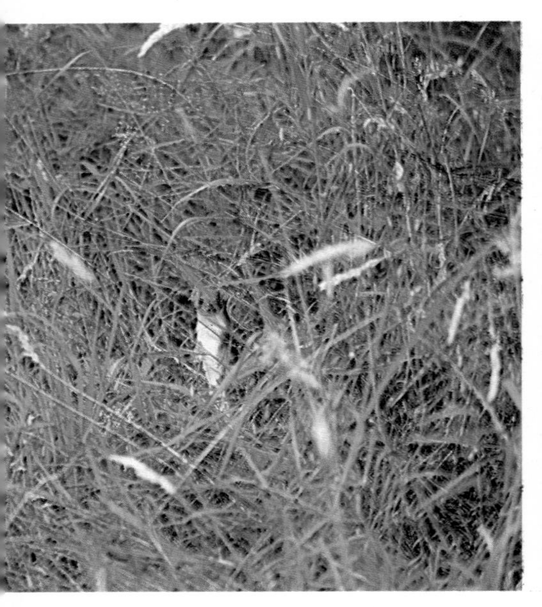

Left *The weasel is Britain's smallest carnivore – so small that it can squeeze down mouse burrows and runs.*

Right *Polecats have been hunted to extinction over most of their range, but are now spreading from their Welsh stronghold.*

unpleasant compared with otters'. The animal has chocolate-brown fur usually with a white patch on the chin, and a fairly bushy tail.

There is always a danger that an animal introduced to a new habitat can upset the balance of nature. This was the worry when American mink escaped from fur ranches and began to colonise waterways. It was feared that streams and ponds would lose all their moorhens and water voles, and that fisheries would be devastated. To date, there is no evidence that the mink has caused anything but local destruction of wildlife, nor is there undue competition with otters. Domestic ducks and chickens, or fish in hatcheries, are at risk if not properly guarded.

Mink hunt on land and in water, concentrating on whatever is abundant, especially fish, birds and mammals. They swim well but dive for only a few seconds. Their main fish prey is eels, which are relatively slow swimmers, and other fish are caught mainly in winter when their movements are sluggish.

POLECAT

The polecat's name comes from the French 'poule-chat' – the cat which eats chickens. As such, it has always been persecuted but it suffered a huge decline in the nineteenth century. The first one exhibited in London Zoo was caught in Regent's Park – then on the edge of London. Once common throughout Britain, mid-Wales became its final refuge, but within the last thirty years it has spread back across the whole of Wales and into the bordering English counties. The polecat is larger than a stoat – about the same size as a mink. Its coat has dark guard hairs with creamy underhairs which show through as it moves. The face is marked with a white patch behind the eye and a white edge to the ear. Identification of polecats is complicated by confusion with escaped ferrets. These animals are domesticated forms of the polecat, but they interbreed with polecats to produce hybrids which are virtually indistinguishable from the wild animal. The polecat lives in a wide variety of habitats, usually where there is extensive undergrowth.

PINE MARTEN

This is the most beautiful of all the slender, weasel-like carnivores. It has a rich chocolate-brown coat with a creamy yellow patch under the throat and a bushy tail. Although stoats and weasels are good tree climbers, the pine marten excels, aided by large, hairy-soled paws with sharp claws for gripping and a bushy tail for balance. It can leap from tree to tree and land on all fours after falling. In Scandinavia the main prey is red squirrels but British pine martens are more terrestrial. The den is as likely to be in a rock crevice or a hole in a bank as in a tree hole or the disused nest of a squirrel or bird. The prey is extremely varied; it includes small birds and mammals, insects, earthworms and, for a carnivore, a large amount of fruit.

Foxes have become one of the more familiar wild animals as they have colonised suburban areas in many parts of Britain.

The pine marten is probably the most elusive of our carnivores. This is partly because of its rarity. It was prized for its fur but was also hunted for sport or as vermin. Not surprisingly, the pine marten went into decline; the last record from near London is of one shot in Epping Forest in 1883. It is now confined to the wilder parts of Scotland, Wales and northern England, but there have been welcome signs of extensions in its range in recent years, and in some areas such as Beinn Eighe National Nature Reserve it is relatively common.

BADGER

The badger is the only carnivore which is relatively easy to find and watch, by virtue of its habit of living in extensive underground setts. The main setts are elaborate tunnel systems – one excavated in Gloucestershire proved to have 310 metres (340 yds) of branching and rejoining tunnels linking thirteen entrances and it had necessitated the removal of twenty-five tonnes of soil.

A pile of soil outside the entrance is a good indication that a burrow was excavated by

One of Britain's largest mammals (up to 0·8m (2½ft)), badgers still survive in most of Britain despite centuries of persecution. They live in large underground burrows (setts) from which they emerge at night.

badgers, but it may not currently be occupied by them. Rabbits and foxes take over empty setts, but clues that badgers are in residence come from scraps of dead ferns or grasses dropped while being brought in for bedding, large five-toed footprints and latrine pits with fresh dung in the vicinity of the setts. Well-worn paths run from the sett to adjoining setts, and to drinking and feeding places. Overturned dry cowpats and conical holes in the pasture show where badgers have been feeding.

Unlike other members of the weasel family, badgers are social animals. The basic unit is the sow and her cubs, and there may be several breeding sows in a group, together with one or two adult boars. These animals share one or more setts within a territory which is defended occasionally by fighting but more often by regularly patrolling the borders and using the prominent latrine pits as 'scent beacons'.

Badgers are almost wholly nocturnal – our other carnivores are more likely to be seen by day, especially if humans take care to conceal themselves. They emerge from their setts after sunset, sometimes earlier in high summer and considerably later in winter, but with a regularity that makes badger-watching a worthwhile pastime.

After emergence, and perhaps following a game or a good scratch, the badgers set out to look for food. Despite their powerful jaws, badgers feed on small fry; they are omnivores and best described as gatherers rather than hunters. Rabbits, mice and voles are often eaten but the preferred food is earthworms. More often than not, badgers eat nothing but earthworms, but the diet is very wide and includes frogs, slugs, beetles, berries, grass, wheat and barley.

Cubs are born in the sett between mid-January and March and they remain below ground for two months. During this period they are active enough to explore the inside of the sett before they eventually come to the surface. At first any noise sends them scuttling below, but they soon gain confidence and start to play boisterous games. Weaning begins at three months and the cubs start to join their elders on foraging trips. They remain with their mother at least until the autumn.

Badgers are a widespread species which, given the vulnerability of their setts, has survived well. Although rare or absent from a few areas such as the flat, easily-flooded East Anglian fens, badgers are common where there are suitable foraging grounds, soil which is easy to dig but will not cave in, and a minimum of disturbance. Following nineteenth-century persecution, their numbers have increased and badgers have adapted to the urban sprawl. However, there have been local declines owing to badger-digging for sport and gassing on sporting estates. Increased road traffic has taken its toll because of the badgers' insistence on using traditional tracks, but provision of special 'underpasses' on new roads has gone some way to solving this problem.

The possibility that badgers transmit tuberculosis to cattle has resulted in an unknown but large number being gassed, and now trapped, especially in the south-west of England. Control measures are confined to very limited areas and the population at large is not affected, but, to date, there is no certainty that killing badgers has reduced the amount of tuberculosis in cattle. Tuberculosis has been identified in badgers and cattle on National

The otter has nearly disappeared from lowland Britain, but is still to be found in more remote parts of Scotland and Northern Ireland. A reintroduction programme has been started recently in Norfolk.

Trust properties, and while the Trust is distressed at the need to control the badgers it reluctantly accepts that it must do so until badgers are proved not to infect cattle.

OTTER

The otter is the most severely threatened of all our carnivores. While other hard-hit species are showing signs of recovery, the otter population has gone into a steep decline. Otters flourished everywhere in Britain until the mid-1950s, except in the polluted, disturbed rivers of industrial regions. Over the next decade there was a sudden slump; otters disappeared from many river systems and the reason was not obvious. Two factors seemed to be the most likely culprits. Rivers and streams are increasingly 'tidied up' through the removal of bankside trees and water vegetation, so destroying the cover needed for breeding and resting, and increased pleasure use of rivers has created enormous disturbance. Otters are now almost confined to the wilder parts of the north and west, and the chances of a recovery seem very slight. Otters are living and breeding on some National Trust properties, where parts of their habitat are left undisturbed by paths being routed to lead the majority of visitors away from their haunts.

Otters are solitary animals; each one inhabits a stretch of water, along a river, lake, stream or sea coast, and tolerates an overlap only with members of the opposite sex. The territory of a dog otter may be over 15 kilometres (9 miles) long and embrace the territories of more than one bitch. The territories are marked with spraints – droppings impregnated with a not-unpleasant-smelling secretion. Spraints are deposited in places which are landmarks to an otter in the water, such as a boulder, the footing of a bridge or in the angle where two streams meet. A well-beaten path may show where an otter habitually leaves or enters the water, perhaps taking a short-cut across a headland.

Otters are adapted for an amphibious life. The ears and nostrils close when submerged and a good set of whiskers assists the eyes when swimming in murky water. Swimming is achieved with vertical beats of the tail, aided by simultaneous thrusts of the paws at slow speed. Yet for all their skill at swimming, otters do not spend long periods in the water and they hunt with frequent short dives lasting less than a minute. Most of their prey is fish and they prefer eels or other slow-moving species.

Cubs are born at any time of the year in a holt under tree roots, in a burrow or in a cave. They first enter the water at about four months and they may have to be dragged in by their mother. Thereafter they take to the water readily and accompany their mother for at least one year.

Although superficially similar to the domestic tabby, the wild cat can be distinguished by its thick, striped, bushy tail.

WILD CAT

Once widespread through mainland Britain, by the nineteenth century the wild cat was limited to the Highlands of Scotland. It is distinguished from the domestic tabby by its large size, blunt bushy tail with cross-striped, never blotched or longitudinally striped, body. Unfortunately, hybrids with domestic cats can make identification difficult.

The wild cat's behaviour is not well known because it is nocturnal and very shy, as well as rare. It feeds mainly on rabbits, followed by rodents, hares and birds. Each cat lives alone in its home range but males (toms) travel in search of a mate in spring. Kittens are born from May to September and become independent in four months.

An often overlooked but common British carnivore is the feral cat – a domestic cat which has gone wild. Feral cats are found in towns and countryside, where they live in small colonies and are often important predators on the local wildlife.

The carnivores have suffered from two main sources: direct persecution, and the destruction of their habitat. The two were linked in the case of the wolf, which was controlled in the Scottish Highlands by the simple process of burning down the forests. Habitat destruction accelerated in the eighteenth and nineteenth centuries, with the massive increase in agriculture and the cutting down of forests for building materials and for charcoal to use in iron smelting. At the same time the heavily keepered shooting estates came into being and carnivores were destroyed wholesale as vermin. Carnivores are no longer persecuted to the same degree but habitats continue to disappear, and the National Trust plays an increasingly important part in the conservation of carnivores by maintaining areas of undisturbed countryside.

A litter of nearly full-grown polecats playfully fighting.

4 *Small Mammals*

Dr PAT A. MORRIS

Most British mammals are small or nocturnal. Many are both and consequently the daytime visitor to National Trust properties is likely to miss what are in fact some of Britain's commonest and most widespread animals. Unless we are prepared, we walk through the lanes and woods unaware that there may be six or eight species of small mammals within a stone's throw of our feet.

Sometimes these elusive creatures give themselves away by making loud noises. Shrews for example are aggressive and when they meet often emit a sharp, high-pitched and ill-tempered burst of squeaks, easily audible from a distance. Moles betray their presence by the piles of earth pushed up from their burrow systems. These are most prominent in pastureland, but can be found in woods though there they are less obvious. Squirrels build large nests in the tree canopy, choosing leafy twigs for construction. In consequence, a squirrel's 'drey' usually has dead leaves visible on the outside in contrast to the similar-sized magpie's nest, which is built of dead twigs and rarely has shrivelled leaves around it.

There are about twenty species of small mammals on mainland Britain (some of them occur on islands too). They form a number of clearly distinct groups.

SHREWS

These are tiny, with a body length of 40–90mm (1½–3½ins). They have very small eyes (pin-head size) and tiny ears. Their fur consists of hair which is the same colour at the tip as the base (if you blow the fur the 'wrong way' the base of the hairs do not appear black as in voles and many mice). Most distinctive of shrews is the long, tapering snout which is constantly on the move. Indeed all shrews are very busy, active creatures, although unfortunately you are more likely to see them dead than alive. There are two reasons for this. They have strong-smelling skin glands that make them distasteful to cats and other predators, so although they are often caught, they are then just played with and left while

The squeaking of shrews is a familiar sound in country lanes, and with patience common shrews (left) *may be glimpsed scurrying through the leaf litter. Water shrews* (right) *look like silver bubbles when submerged, but can often be seen far from water.*

their killer goes in search of something more palatable. Dead shrews are also seen lying on paths in the summer, giving rise to the legend that they are killed by loud noises like thunder claps. In fact it is usually 'old age' that is to blame. Few shrews live to be much more than a year old and are so worn out after the breeding season that all the adults die before the autumn, leaving the juveniles to survive until the following year. It is these dead shrews that are often picked up by walkers, puzzled by the lack of an obvious cause of death.

There are three mainland species of shrews: the water shrew is the biggest, and is distinctly coloured black above and silvery white below. Despite its name it is often found far from water. Common and pigmy shrews are shades of brown; the latter is Britain's smallest mammal, being only about 50mm (2ins) long. These two species are best told apart by the relative length of the tail. It is about three-quarters the head and body length in the pigmy shrew and half the head and body length in the common shrew. All three species are most abundant in damp areas of woodland and grassland.

MICE

Most mice have big ears, big eyes and a long tail. House mice are greyish brown all over and tend not to be found far from buildings. Wood mice (also called long-tailed field mice) are very elegant-looking, delicate creatures with a pure white underside and sandy brown back. Their huge ears and bulging black eyes are important adaptations to their strictly nocturnal life. Although they are most common in woodland and hedgerows, they also venture out on to sand dunes and shingle beaches; but only under cover of darkness. In parts of southern England a very similar species is found: the yellow-necked mouse. As its name suggests this pretty creature has a broad yellow collar round the throat, almost like a little harness linking the two front legs. However, this is usually only visible if you find one dead. They are good climbers, readily exploring bushes and trees far above the ground, and coming into outbuildings and attics, especially in the autumn.

Harvest mice are small – 60mm ($2\frac{1}{2}$ins) plus tail – and a golden yellow colour. They have rather small ears for mice, but are very distinctive. They behave almost like minute monkeys, scrambling about among the twigs of hedges and in the stalks of tall, stiff-stemmed vegetation. Though popularly associated with corn they cannot live in it permanently because, of course, the crop is reaped annually. Similarly, tall grass dies back in the autumn, at which time harvest mice often can be seen and they then spend more time at ground level. They are surprisingly common in wet areas, where they climb about and nest in the tall reed stems and tussocks of marsh grass. Their nest, a closely woven ball about the size of a cricket ball, is quite distinctive and often very obvious in winter when the vegetation has died back.

Dormice are the only mice that have a furry tail. They are strictly nocturnal and spend most of their time up in trees and bushes or asleep out of sight in a nest. The so-called

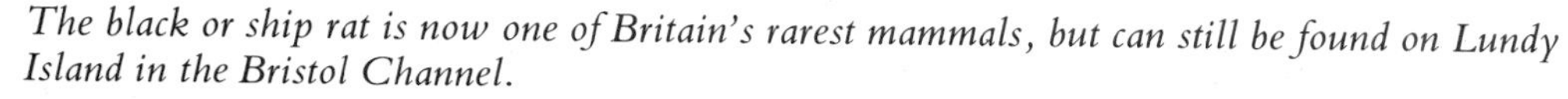
The black or ship rat is now one of Britain's rarest mammals, but can still be found on Lundy Island in the Bristol Channel.

common dormouse is now quite rare and anyway confined to southern and midland counties. This is the animal immortalised in the Mad Hatter's tea-party, and when *Alice in Wonderland* was written dormice were popular as children's pets. It builds a nest often in among roots or in the base of a coppiced tree. The common dormouse is a bright orange yellow all over, whereas the fat or edible dormouse is grey and much bigger, being the size of a small squirrel. It was introduced from the continent of Europe at the turn of the century and is found mainly in the Chilterns, where it lives in woodlands and often enters houses to hibernate.

RATS

Most people would rather not know about rats. They are large, destructive rodents, usually associated with buildings, though their burrows are often evident in hedgerows near by. Brown rats are very widespread, but black rats are now confined to certain cities and dockland areas where they live in old buildings. The black rat is now one of Europe's rarest mammals and could be considered endangered!

The brown rat was not known in Europe till 1553 and in Britain till the eighteenth century, when it was known as the Hanoverian rat. It can have litters of up to twenty, though the average size is eight to nine. The young rats are weaned at three weeks and can be sexually active by the age of two months. The black rat, on the other hand, is slower to mature, which is one of the reasons it has been displaced in the British Isles by the brown rat.

The popular image of the Trust is as a conservator of old houses, and where better to find house mice and thriving rat populations? In practice, the situation is not that simple. Once, such large, rambling houses with open kitchens, big larders and food stores would have been heavily infested. Today they are not, because rats and mice prefer warm places and they must have a regular supply of food. Many of the Trust's large houses are unsuitable in that they are not fully lived in (and may even be closed up for the winter months) so that food is no longer plentifully available throughout the year.

VOLES

Bank voles and field voles (also called the short-tailed vole) are chubby, snub-nosed creatures with smallish eyes, small ears and a small tail. The former is a chestnut brown on the back and characteristic of woodland and hedgerow, while the latter is the typical small mammal of old meadows and long grass. It is usually grey or yellowy brown on the back and grey underneath, with a very short pink tail. Both species are important food for owls, other birds of prey and predatory mammals. The runs of voles, less than 25mm (1in) in diameter, can be found among grass roots and leaf litter.

Bank voles and wood mice are particularly fond of hazel-nuts and a careful search in woodland usually reveals plenty of empty shells with a neat hole gnawed in one side. It is even possible to tell which was responsible; when mice chew hazel-nuts they make little tooth marks in the brown shell-surface all round the edge of the hole. Gnawing by bank voles leaves a neater hole with a sharp outer edge. Vacated birds' nests are often used as 'dining tables'. Dormice leave gnawed nuts lying about too, but these are seen less often and need careful examination to distinguish them from wood-mouse-chewed specimens.

The water vole is a large, stocky-looking animal, the size of a rat. The original 'Ratty' of *The Wind in the Willows*, it is not actually a rat at all. Water voles normally live at the edge of lakes and rivers where they make distinctive burrows in the bank at water level. They feed on waterside plants, often leaving a lot of tell-tale chewed leaves and stems about, with perhaps a sprinkling of black oval droppings too. Water voles are much more active in daytime than most small mammals and so the quiet, patient watcher may well be rewarded by the sight of one or more of these inoffensive creatures swimming, diving and going about their normal business. Practically any of the National Trust properties with undisturbed rivers or lakes are likely to have water voles present.

SQUIRRELS

Everyone knows what squirrels look like, but there is often confusion between the two species. In late summer grey squirrels often have quite brown fur and may be mistaken for red squirrels by people who are not familiar with the latter. The red squirrel is a bright chestnut orange in summer, whereas the brownest grey squirrel is at best a brownish khaki-colour. In summer our red squirrel has a bleached, creamy white tail (a feature which distinguishes it from its continental cousins), whereas the grey squirrel always has tail hairs that are banded with black, brown and white, giving a grizzled appearance. In winter, grey squirrels are a smart silver grey on the back but reds are chocolate brown with a plain brown tail and perky ear tufts (which are never found in the grey).

Both squirrels are characteristic woodland animals (though the grey is more willing to frequent gardens, hedgerows and farmland trees). Both are active in daytime and are thus among the most likely species to be encountered on National Trust properties though the red is increasingly rare outside Scotland. Both will frequently become very tame, taking food from visitors and raiding litter bins.

RABBITS AND HARES

These too are distinctive animals, but often confused with each other. Rabbits usually live in social groups, dig burrows and do not often venture far from cover. You may see them – given away by the quick bobbing of their white tails as they run – on the edge of deer parks, or woods in National Trust properties, especially late in the afternoon when they come out to begin feeding. Hares like to live out in the middle of wide open spaces. Brown hares inhabit grassland and farmland; mountain hares live on moorland and upland areas, mostly in Scotland but also as introductions in the Pennines, where they occur on the newly acquired Trust land of Kinder Scout in the Peak District. They turn white in winter, except in Ireland.

Rabbits have a well-deserved reputation as prolific breeders. Their gestation period is about a month, and the average size of litter is five to six. The does can conceive again immediately after giving birth, and have between five and seven litters a year. The young are blind and naked when they are born, though they leave the nest at three weeks and are weaned at four weeks. The hare has a much smaller litter – usually only two. The gestation period is six weeks and the young are fully formed and active from birth.

Brown hares have a grizzled brown fur and decidedly big orange-brown feet. They are surprisingly large, lanky-looking creatures as they lope along, and they have big staring eyes. Rabbits are smaller, with shorter ears, and are greyish-fawn with a buff-coloured patch on the nape of the neck; they lack the prominent black ear tips seen in hares. The presence of rabbits is often shown by tell-tale signs such as heaps of dry spherical droppings, but hares do not advertise themselves much and are normally hard to see as they lie still during the day and are mainly active in the evening and early morning.

MOLES

Although the mole is one of our most widespread mammals (but absent from Ireland like the common shrew, field vole and some others), it is not often seen alive. Dead ones are sometimes found and are unmistakable with their all-black fur, big pink hands and almost invisible eyes. Moles feed mainly on earthworms, which they paralyse with a bite and keep in temporary stores in their underground tunnels. Most people know moles from their molehills, which are called tumps, but they also build larger, breeding 'fortresses'.

Left *The mole is rarely seen above ground as it spends most of its life in subterranean burrows which it excavates with its powerful, spadelike forepaws.*

Right *Hedgehogs are probably Britain's most popular mammal, but unfortunately many are killed on suburban roads.*

Brownsea Island in Poole Harbour, Dorset, has a thriving population of red squirrels. The tails of British red squirrels tend to bleach in summer.

HEDGEHOGS

The hedgehog is probably Britain's most unmistakable and best-known mammal, but it is not often seen alive, being nocturnal in its habits. At dusk it can sometimes be heard as it forages in dense undergrowth. For such a small creature it makes an amazing amount of noise. Parkland, hedgerows and suburban gardens are among the hedgehog's favourite habitats.

Most small mammals are very numerous. Some, like the dormouse, are rather difficult to study so that we do not really know whether they are abundant or scarce. None of them is rare enough to be threatened with imminent extinction, though the populations of brown hares and red squirrels have seriously declined in many areas and give cause for concern. The hare seems to be a victim of modern farming methods. It likes big open fields, but these tend to be cultivated and sprayed just at the time when baby hares (called leverets) are at their most vulnerable. Worse, the use of selective herbicides to kill weeds in farm crops may deprive hares of vital food species. Whatever the cause, hares are becoming more scarce in many areas and population levels are not helped by the continuation of hare shooting in many places.

Red squirrels, once widespread over most of lowland Britain, are now extinct throughout practically all of the south and east and most of Wales. They have been replaced by introduced grey squirrels, whose spread continues year by year. Once greys become established, reds seem unable to recolonise the land they once occupied. Whatever the reasons for the red squirrel's retreat and diminishing distribution (and there is still much debate on this, but little research and few real answers), we are witnessing the gradual decline of one of Britain's prettiest native mammals. Despite the damage it does to forestry interests (just like the grey), it would be a pity if it became extinct. The National Trusts have an important role to play here. As custodians of much of Britain's woodland, the Trusts safeguard essential habitat for this species. The red squirrel has now also been given legal protection; but neither may be enough if the squirrel's decline is due to some, as yet, unknown factor. In England the Trust also owns the red squirrel's only remaining strongholds in the south: Brownsea Island and parts of the Isle of Wight.

Some National Trust properties provide secure and undisturbed sites which are ideal for long-term ecological research projects. Cambridge University, for example, has carried out regular monitoring of small mammal populations at Wicken Fen.

Despite their cumbersome appearance hedgehogs are agile climbers and can swim well. However, they frequently drown when they fall into water butts and steep-sided ponds.

Brown hare populations have declined in many areas due to changing methods of agriculture, which leave little suitable habitat.

The National Trust also owns a number of islands that are the homes of interesting isolated mammal populations. On the Farne Islands, for example, there are black rabbits. On Lundy there are both brown and black rats. It is unusual to find these two species together and the latter living outdoors in wild habitats. Yet on Rat Island (at the southern tip of Lundy) black rats found a home in the cliff-top vegetation, exposed to gales and sea spray. The island of St Kilda (owned by the National Trust for Scotland) is one of the most remote parts of Britain, the outermost of the Outer Hebrides. It is so remote that land animals could not get there unaided, and yet there are wood mice present! These must have arrived by boat, carried there accidentally perhaps with a cargo of supplies. They may even have arrived with the Vikings, but have certainly been there long enough to have evolved a separate and distinctive island form. It is twice the size of the mainland wood mouse, with much longer fur and is found nowhere else. Once St Kilda had its own distinctive race of house mouse, but this died out after 1930 when the island's human population was evacuated.

The Trust's efforts to popularise our natural heritage and promote visits to the countryside have meant inevitably that very large numbers of people now visit places that would otherwise have been left alone. For the most part, the trampling of human feet and generally increased disturbance are not a serious threat to small mammals. However, in some localities there may be problems and everywhere there is trouble caused by litter. Discarded bottles, left by untidy picnickers, are a particular hazard to smaller mammals.

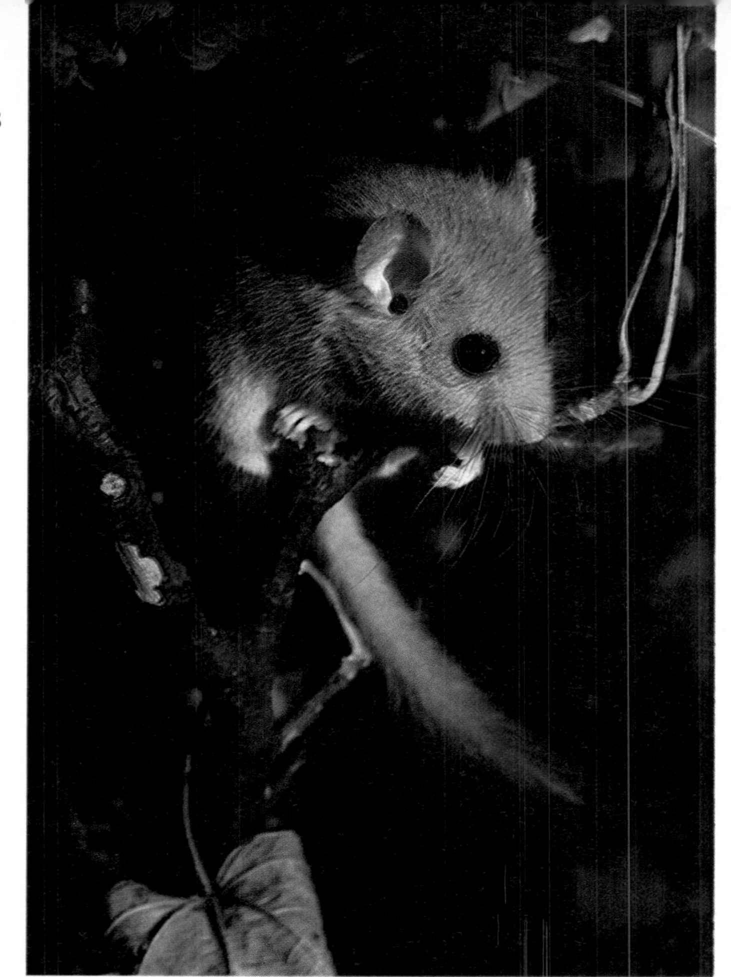

Dormice, once popular as children's pets, are rarely seen. They are nocturnal and usually live in coppiced woodlands with wild honeysuckle.

Millions of bottles are left lying about the countryside every year. Mice, voles and shrews enter them in search of food or in the course of exploration, and then they cannot get out again. They are trapped and starve to death or die of cold. Sometimes a single bottle will claim twenty or more victims, including less common species such as the dormouse and water shrew. Discarded 'ring-pull' drink cans kill animals in the same way. Every effort should be made to stop people leaving any sort of litter, but the thoughtless abandoning of bottles is particularly regrettable (and even worse is the bottle carefully hidden away in a bush 'out of harm's way'!)

There is one small way in which litter might be said to be beneficial. On certain popular mountain tops there is practically no food save for apple cores, orange peel and other debris left by visitors. This source of nourishment may be a factor enabling wood mice, for example, to live around the summit of Snowdon in Wales on an otherwise bare mountain top. However, people leave bottles about up there too – and these kill the mice. There is really only one thing to be said about litter: take it home with you.

5 Bats

Dr R. E. STEBBINGS

Bats are the only mammals that fly and fifteen species are found in Britain. This represents nearly one-third of our native mammal fauna. All species are found in southern England, while only four live in northern Scotland; bats can be found in or around all National Trust holdings.

Some National Trust properties which are used for roosting are extremely important, especially for nursery roosts which contain bats that gather from many hundreds of square miles. One building in west Wales contains a nursery roost of the endangered greater horseshoe bat and is one of only six such sites in Britain.

Unlike many other mammals, bats are heterothermic, that is they have widely varying body temperatures. In flight their body temperature is 40–2°C (much higher than man's 37°C), but after landing their temperature rapidly falls by about 10°C, while they digest the food they have eaten on the wing, and it may later be dropped to the surrounding temperature in order to conserve energy. In flight a bat's pulse rate goes up to 1,000/minute (man's is around 75/minute).

Stackpole in Pembrokeshire, which has breeding colonies of four species of bats, including the greater horseshoe.

Common long-eared bats frequently roost in buildings. Until recently woodworm treatments were often lethal to bats, but less toxic preservatives have now come on to the market.

Bats have a low rate of reproduction, normally producing a single baby (occasionally twins) each year, and not starting to breed until they are two years old. Although they mate in autumn and winter, fertilisation is delayed until spring, and then development proceeds at varying rates depending on the availability of food. Insects form their main diet and cold weather prevents the bats from feeding, so that pregnant bats become torpid in bad weather and gestation is therefore lengthened. The babies are born in June or July and they can fly after about three weeks, growing to full size within two months. Following birth, the baby is raised with great care by its mother and suckled frequently, both by day and night. Young adult females often breed only in alternate years, probably because it is so energy demanding.

Bats normally do not use sight to navigate by, but a form of sonar or echo-location, using high-frequency sounds of up to 120,000 Hz (cycles per second) compared with man's maximum hearing range of 20,000 Hz.

Pipistrelle bats, our smallest and by far the most abundant species, may each consume up to 3,500 insects nightly. As winter approaches bats lay down food reserves in their bodies in the form of fat, resulting in weight increases of up to one-third.

Hibernation is a series of varying periods of torpor broken by arousals when flight often takes place. Bats select their hibernation site very carefully, based mainly on temperature. Each species has its own preferred temperature range. In early winter warm sites are chosen, but as food reserves become depleted they move to cooler roosts which may be a few yards or many miles away. Fat bats also choose warmer sites than thin bats, and because of these exacting requirements bats may be found occupying a wide variety of secluded roosts.

There are three major types of roost. Buildings of all kinds, including ancient monuments, churches, fortifications, follies, farm and domestic houses, are the most important in summer. Cave-like places, including natural caves and rock clefts, mines, cellars, lime kilns, ice houses, underground fortifications and tunnels of all kinds (railway, canal, grottoes, servant's and service) are used mostly for hibernation. Hollow trees in hedgerows and woodlands are used throughout the year. All these differing roost types are found on National Trust properties and some are specially protected for bats.

Some species are confined almost exclusively to hollow trees (for example, noctule and Bechstein's). However, some that were totally dependent on caves (for example, the greater and lesser horseshoe bats) have now adapted to using various parts of buildings. Others use hollow walls, roof spaces and the gaps behind weather boarding or hanging tiles and above soffits. In many larger buildings, including stately homes, bats can be found in almost any cavity.

Different types of roosting site may be used by an individual bat during the year. For instance, a brown long-eared may be found in a cave in December, a hollow tree in April and the roof of a house in June. Consequently a roost is not usually occupied throughout the year, but the same site tends to be used by the same colony at the same season each year. The largest colonies of bats (in Britain they rarely exceed 500) are found in the period June–August. Pregnant females all congregate, often from an area of several hundred square miles, in one favoured nursery roost. These roosts are vital to the continued survival of whole populations, but are extremely vulnerable to catastrophes such as a fire, which can destroy the breeding potential of the species over a large area. When the young are weaned the adult females usually leave, followed later by the young. Adults begin leaving the nursery in early August, and in almost every case all bats have gone by the end of the month. In large buildings with complex roofs, bats may move from one part to another and live within the same building throughout the year. Large buildings may house several colonies of differing species; one building in Dorset had at least seven breeding species.

Surveys have revealed that a large proportion of buildings are prospected by bats and many colonies occur in buildings less than fifteen years old. Colonies have even been known to move into new buildings before they are completed.

The ideal habitat for bats to feed in varies from species to species but most need sheltered areas with plentiful insects. The best areas are parkland with permanent, unimproved pasture, woodland and river valleys with slow-moving rivers, meadows and marshland. Many National Trust properties have such ideal habitats, and hollow trees in these areas provide the major roosting places for bats. Poor areas include exposed moorland, intensively cropped arable farmland with few hedgerows, and large monoculture plantations. These areas often lack both roost sites and large numbers of insects.

CONSERVATION PROBLEMS AND SOLUTIONS

Originally bats were animals of woodland and river valleys, roosting in hollow trees and caves. With the clearing of Britain's woodland cover they adapted to living in buildings. Nothing is known of population changes in historic times but modern declines have been severe. The greater horseshoe has declined by over 98 per cent in a century and bats which until recently were 'common' appeared to have declined by about 50 per cent in three years. The main causes of these declines have been loss of roosts, loss of feeding habitat and food, bad weather at critical times and use of insecticides to kill 'woodworm' in buildings.

Roosts in natural caves in Britain are confined to a few areas mostly in limestone, but bats quickly colonise man-made mines and other underground buildings, including disused railway tunnels. However, in the past thirty years many such sites have become unusable because of entrances blocked for 'safety' or by rubbish tipping and accidental disturbances, for instance by industrial archaeologists and cavers. Hibernating bats have also been killed by vandals. About thirty-five greater horseshoes were recently killed in a south Devon cave by people using sticks and fireworks, and in 1980–1, ninety-six bats were killed in a Suffolk tunnel through being set on fire with matches and cigarette lighters, shot with airguns and crushed with sticks and stones.

The number of roost sites in buildings is reduced when access holes, such as ventilators, are blocked and cavity walls are filled for insulation. Hibernating bats may be engulfed by foam during this latter process. The modern need for security around buildings has resulted in former roost sites being denied to bats, for instance cellar ventilator holes and doors being closed. Retiling and underfelting roofs of old buildings often results in the exclusion of colonies. Remedial timber treatment is probably the greatest threat. Over 200,000 buildings are treated annually with chemicals, many of which are lethal to bats. 'Woodworm' killers are extremely persistent within buildings and even if bats are not present during treatment, when they return they will pick up poison by inhalation of vapour and contact with treated surfaces. These chemicals can pass through the skin but most are swallowed when the bats groom themselves.

Trees with holes are important for many creatures, including bats. But such trees are often regarded as dangerous and are felled or 'tidied'. Trees become of most value to wildlife when they are regarded by foresters as past maturity. Also riverside trees, many of which are pollarded and hollow, are removed for convenience during canalisation. Extensive lakes with mature parkland, such as that at Stourhead, provide ideal bat habitats.

Habitat destruction may affect bats' food supply. Elimination of permanent unimproved pasture and hedgerows, felling of woodland and canalisation of rivers has greatly

A cluster of hibernating greater horseshoe bats.

reduced the number and variety of night-flying insects and the sheltered places in which bats can feed on them. Cockchafer beetles are now far fewer but are vital to the large bats after hibernation. They spend about three years in the soil feeding on roots and sometimes are regarded as grassland pests. Pasture may be sprayed specifically to kill cockchafers.

A change in farm practice from hay-making to cutting grass for silage has meant that many insects which mature when plants flower are now killed before they breed. Agricultural and forestry insecticides may result in traces of chemicals in insects which can then accumulate in the bats feeding on them. Large accumulations may directly kill some bats and small amounts may reduce breeding success.

Although the threats to bats are very severe and undoubtedly populations have declined rapidly, some positive measures are now being implemented. Bats will quickly adopt suitable new sites for roosting. During the course of foraging, possible roosts are investigated. Long-eared bats, which glean food off foliage or branches, find new sites more quickly than other species which catch flying insects or feed on the ground. Bat roost boxes placed on trees or buildings can attract bats, especially if placed in areas lacking roosts and close to feeding grounds. These boxes simulate tree holes and can be used by large numbers of bats of most species. Although there is always a chance a box might be used irrespective of its setting, one can improve the likelihood by placing them in areas where there are few alternative roost sites. Bat (and bird) boxes can be seen in many National Trust woods both in deciduous and conifer plantations.

Cold cellars can be made suitable for hibernating bats by providing iron gratings over external ventilation holes, and the cellars in several National Trust houses have been set aside as bat sanctuaries. However, as bats should not be disturbed, they are not normally available for visitors to see.

An increasing problem faced by the National Trust, as with other owners of valuable property, is that of security. Alarms are sometimes activated by bats flying around rooms and corridors at night. Repositioning and adjusting the sensitivity of sensors or even changing the system will cure the trouble.

Perhaps the most important conflict in conservation facing the National Trust is the need to preserve their buildings using chemicals which kill wood-boring beetles and fungi, but often these chemicals are also lethal to bats. The problem is being investigated but until recently there was no solution to the problem apart from replacement of timbers. Now the development of permethrin, which is far less toxic to bats but an effective woodworm treatment, gives considerable hope for the future.

SPECIES FOUND ON NATIONAL TRUST PROPERTIES

All fifteen species of British bats are found on National Trust properties, including the rarest British mammal, the mouse-eared bat. This is virtually extinct; by 1982 there were only two known surviving individuals, both males. They hibernate in a specially protected National Trust tunnel in southern England.

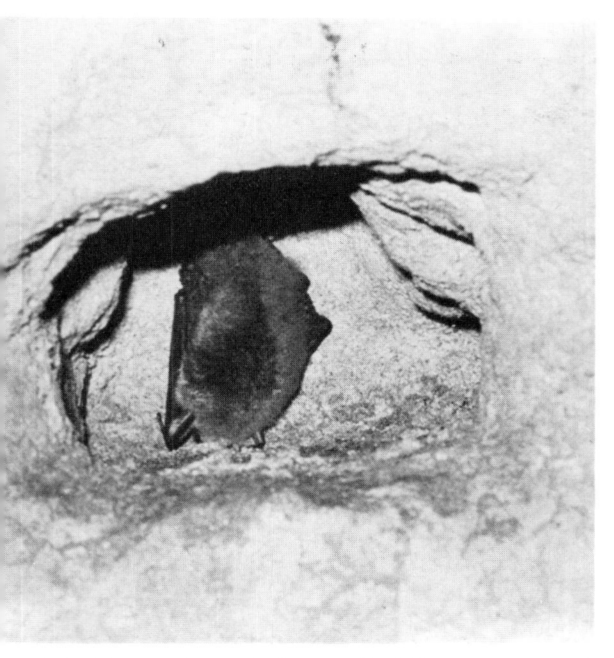

The Natterer's bat (left) *usually hibernates in crevices, whereas the greater horseshoe bat* (right) *hangs free. This greater horseshoe has a metal band on its wing which is used for identification purposes – marking of bats can only be carried out under licence.*

Other National Trust properties in the south-west are very important for maintaining the survival of another endangered species, the greater horseshoe bat. In Britain this species now numbers only about 2,200 bats and is divided among six main colonies. A century ago there were an estimated quarter of a million bats throughout southern Britain. Many of the colonies were deliberately killed. Casual or deliberate disturbance during hibernation causes the bats to wake and use up food reserves, and several such disturbances in one winter will result in bats dying of starvation.

Greater horseshoe bats are very difficult to see because they fly close to the ground or among foliage, usually when it is quite dark. The tiny lesser horseshoe bat is equally difficult to see but is not as rare. They form nursery colonies in buildings, some in National Trust properties in western Britain; they also roost in cellars, mines and other underground sites. Worldwide the lesser horseshoe bat is an endangered species, and in Britain colonies are apparently declining, but the reasons are unknown.

Perhaps the most conspicuous bats are the large, narrow-winged, noctule bats, and similar but smaller Leisler's or lesser noctule bats. They can often be seen flying fast and high over trees, making sudden dives when they chase insects.

Another large bat, with broader wings, which may be seen feeding below tree-top level over parkland is the serotine. Many National Trust houses shelter colonies which are often resident throughout the year.

The mouse-eared bat is Britain's rarest species. At the time of writing only two males remained.

Pipistrelle nursery colonies are mostly found in buildings and sometimes contain several hundred bats. Pipistrelles chatter noisily all day in warm weather but especially as they begin to emerge about twenty minutes after sunset. Most National Trust buildings will be visited by this species and many contain colonies. However, even this once very common bat has suffered a rapid and severe decline which is probably due, at least in part, to cold, wet weather in June and July reducing the number of insects. Chemical remedial treatment of timber in roofs may have killed or injured many.

Long-eared bats, particularly the abundant brown long-eared, will also be found in many National Trust buildings and in any associated wooded or parkland area. They form small colonies which emerge late in the evening when it is dark and feed primarily on moths caught above grassland or around trees.

The other six species are all found in National Trust areas, but they tend to be inconspicuous. Daubenton's bat, Natterer's bat, the whiskered bat and Brandt's bat are widespread, but Bechstein's bat and the barbastelle bat are very rare.

All species of bats were protected in the Wildlife and Countryside Act 1981. It is now an offence intentionally to kill, injure or handle a wild bat of any species in Britain, or to disturb a bat when it is roosting. It is also illegal to damage, destroy or obstruct access to any place that a bat uses for shelter or protection; this applies even to domestic buildings. If bats are not wanted or if a building is to be modified in any way that would affect a bat roost, then the Nature Conservancy Council must be notified so that it can provide appropriate advice. The aim of this part of the legislation is to ensure that bats are not accidentally killed.

National Trust properties, both land and buildings, constitute a most valuable resource helping the survival of many of our bats, including two highly endangered species, the greater horseshoe and mouse-eared bats.

Opposite *Although Bechstein's bats usually roost in hollow trees, occasionally they can be found around barns and old buildings.*

6 *Sea Mammals*

ROBERT BURTON

With a group of animals as mobile as the marine mammals it is no easy matter to decide which species should be regarded as regular inhabitants of these islands. For instance, the white whale and narwhal, the walrus and the bearded seal, all of which live in Arctic seas, have wandered down to these shores on occasion. The only record of the pigmy sperm whale for the British Isles is a single animal stranded on the coast of County Clare, Eire, in 1966; but does this indicate a real rarity, or is this whale a more frequent visitor without showing itself?

Recording the occurrence of cetaceans – whales, dolphins and porpoises – has always been hampered by the difficulty of field identification. In the same way as ornithologists once believed it was necessary to shoot a bird in order to confirm its identity, so cetologists have relied on carcass records from the whaling industry and on the records of strandings kept by the British Museum (Natural History) since 1913. Sight records of rare birds became acceptable a long time ago, and a growing band of 'whale watchers' is now gaining experience in identifying live cetaceans. The Cetacean Group of the Mammal Society, for instance, was set up in 1973 and has already made some analyses of cetacean distribution. From them we know that, overall, cetaceans are likely to be spotted in the late summer, especially August and September, and off the northern and western coasts of the British Isles. The Mammal Society has published a *Guide to Identification of Cetaceans in British Waters*, by P. G. H. Evans, but identification is still by no means easy from the glimpse of a whale or dolphin surfacing among the waves.

Seals pose fewer problems of identification. There are only two resident species – the common seal and the grey seal – although it is possible even for experts to confuse the two in difficult conditions. The ringed seal from the Arctic may go unnoticed because of its similarity to the common seal, but the other rarer seals are distinctive and they are sometimes very tame, as was the walrus that swam up the river Ouse in 1981. The resident species have regular haunts which make them among the easiest mammals to see. While they are swimming in the sea below seals take no notice of watchers on cliffs, and in some places they have become accustomed to boats, which can approach them quite closely as they lie on the rocks. When the seals eventually enter the water they will probably surface near the boat to take a closer look themselves before swimming away.

CETACEANS

Cetaceans live underwater and with nostrils – the blow-holes – on the top of the head they can break surface to breathe with virtually none of the body exposed. The larger whales give away their presence by the blow – the column of vapour that jets into the air at each explosive expiration of breath. The smaller species attract attention when they are travelling at speed and leaping clear of the sea every few metres, an action called porpoising.

When cetaceans are travelling, they usually move steadily in a straight line and surface at intervals to breathe; but when feeding they stay in roughly the same place, breathing several times in quick succession, then disappearing while they chase their prey.

Twenty-four species of whales and dolphins have been recorded in British waters.

A stranded common dolphin. Since the turn of the century detailed records of all cetaceans stranded in Britain have been maintained by the British Museum (Natural History).

There is little evidence that any have resident populations, although this may be true for some of the dolphins.

BALEEN WHALES

These are whales which live near the surface of the sea, where they find swarms of fish and other marine animals. They feed by scooping animals from the water by means of rows of baleen, or whalebone plates, hanging from the roof of the mouth. The baleen forms a sort of net which lets the water run out and leaves the prey animals to be swallowed by the whale.

The large baleen whales are most likely to be seen around British coasts while they are migrating between their northern feeding grounds and southern, warm-water, breeding grounds. They move northwards in spring, passing around the western coasts of the British Isles and carrying on up the Norwegian coast. The return trip is made in later summer or autumn. They follow a route along the edge of the continental shelf, well out to sea, but local abundance of fish and plankton may bring them inshore and within easy viewing distance, especially at headlands or islands where eddy currents cause concentrations of marine organisms.

The fin whale is the second largest species found in British waters, growing to a length of 24m (79ft). In recent years whale watching has become popular, especially in the United States, but whales can also be observed in British waters particularly off Scotland and Northern Ireland.

The right whale is a 'skimmer'; it swims through swarms of prey with its mouth open and a current of water continually passes through its baleen. Other baleen whales, such as blue, fin, minke and humpback, are 'gulpers'; they take a mouthful of water, close the mouth, then squeeze the water through the baleen. The sei whale is both a 'gulper' and a 'skimmer'.

Little is known about the diet of baleen whales in British waters. In general, 'skimmers' feed on swarms of tiny crustaceans and 'gulpers' feed mainly on larger animals such as fish. Minke whales, for instance, join the flocks of sea-birds that gather at shoals of sand-eels and they are also attracted to concentrations of herring and mackerel. The fin whale was called the 'herring hog' by British whalers.

Of the six British species of baleen whales, the two most likely to be seen are the fin whale and the minke whale. The numbers of fin whales have declined considerably during this century and recent sightings have been made only in waters off the west of Ireland and the Hebrides, where they appear to follow the edge of the continental shelf. Minke whales are more numerous and are seen between June and October.

In trying to distinguish between fin and minke whales – if a good, close view can be obtained – the characteristics to be examined are the narrow head, in profile, of the fin whale, compared with the deep 'chin' of the minke whale, the white patch on the flipper of the minke whale and the white lower *right* jaw of the fin whale.

Fulmars feeding on the carcass of a dead fin whale. Although once hunted, fin whales are now fully protected in British waters.

Other baleen whales are extremely unlikely to be seen, but should you be lucky enough to see one at close enough quarters to make identification possible, it is worth keeping detailed notes. Records would need to be very detailed to be accepted by the Mammal Society or the British Museum (Natural History). Since the Middle Ages cetaceans have been 'Royal Fish' and any found stranded in England or Wales should be reported to the coastguards, who, in their turn, notify the British Museum (Natural History). Humpbacks are renowned for leaping out of the water, but the trait is not confined to this species and a sight of the long flipper is needed for confirmation. The blue whale is recognised by its very large size and the right whale can be identified by its double blow and the absence of a dorsal fin.

The right whale has been hunted almost to extinction and despite thirty-five years of total protection there is little sign of recovery. Exploitation of other baleen whales has continued in the north Atlantic at ever diminishing levels as stocks have dwindled. It is to be hoped that the total ban on commercial whaling proposed by the International Whaling Commission in 1982 will be introduced, and that it comes in time to allow these species to recover to anything like their previous numbers. We cannot even guess at their former numbers in British waters, but they were certainly once very much more abundant, and probably frequently to be seen, even close to shore.

TOOTHED WHALES

In contrast to the baleen whales, the toothed whales such as porpoises and dolphins are still a common sight in British waters, and they are more abundant off western and northern coasts. However, even their numbers are considerably fewer than they were a century ago. Their habit of associating in schools makes them easier to spot and the

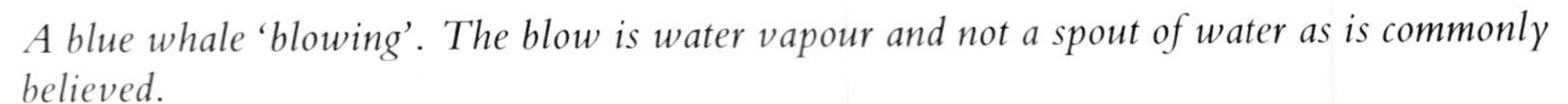
A blue whale 'blowing'. The blow is water vapour and not a spout of water as is commonly believed.

observer may be treated to the marvellous sight of a school of dolphins at play; leaping, rolling, waving flippers and flukes, sometimes with several individuals moving in unison. They will even change course to frolic around boats.

The size of schools is usually less than a dozen but groups of a hundred or more have been recorded. There is some indication of regular migrations, which may be either between warmer and cooler latitudes or from deep to shallow water. There are also correlations between the appearance of dolphins and the shoaling of fishes. For instance, bottle-nosed dolphins, which usually feed on fish which live on the floor of the ocean, enter Scottish estuaries at the time of the salmon run.

The harbour or common porpoise, which is recognised by its small size and low triangular fin, is the most frequently seen of British cetaceans partly, perhaps, because it is an inshore species, and it is one of the few regularly to be found in the North Sea. Common, bottle-nosed, white-beaked, white-sided and Risso's dolphins are the other common species; they are identified mainly by the pattern of colouring and shape of the head.

The killer whale, which is a large dolphin, is often seen around the British Isles and is immediately identifiable by the tall fin of the adult males. Its main food is fish; despite popular belief, other dolphins and seals are relatively rarely eaten. There are surprisingly few records of killer whales at the large seal colonies of the Farne Islands and elsewhere, and fewer accounts of attacks on seals. Even then, the victim sometimes appears to be left uneaten.

Of the larger toothed whales, the sperm whale is not often seen as it usually keeps to deep water; the famed white Moby Dick was a sperm whale. The long-finned pilot whale visits south-west England and northern Scotland especially in winter, mackerel being the attraction. It strands quite frequently and the deliberate driving of schools ashore to slaughter them was formerly practised in Scotland, as it still is in the Faeroe Islands.

Why pilot whales and other cetaceans become stranded is still largely a mystery. Ultimately they are caught by the falling tide but it is not known why they come into shallow water. It may be that the sonar systems of one or more 'senior' animals who lead the school have been damaged by disease or parasites. When the school comes towards the shore, perhaps following fish, the leaders do not notice the shelving bottom and the remainder follow them blindly until it is too late.

Although records are scant, there is evidence that the numbers of dolphins are declining. There are three probable reasons: increased shipping traffic, accumulation of pollutants in the body and dwindling fish stocks.

SEALS

The habits and movements of the grey seal and common seal while out to sea are almost a complete mystery, but for much of the year they can be seen in the vicinity of either

Overleaf *Common seals hauled out on a sand bar.*

their breeding grounds or regular resting places on shore. The grey seal breeds from Cornwall, clockwise around to Norfolk, with the largest aggregations in the Outer Hebrides, Orkney, Shetland and the Farne Islands. Small colonies breed in inaccessible caves around the Cornish coast, but the seals can be seen in the water from many clifftop vantage points, as in the Trust's properties of Boscastle Harbour, Zennor, Godrevy Point and around the Lizard.

British common seals are more restricted and breed from Argyll, north around Scotland and down the east coast to Essex, with the largest concentration, around 6,000, in the Wash. Grey seals prefer exposed rocky coasts and breed on islands and in caves, but occasionally on sand, as on Scroby Sands, Norfolk. Sightings of common seals are more likely in shallow, sheltered water of deep bays and archipelagos, often in view of the shore. They can be seen lying on sandbanks from the National Trust property at Blakeney in Norfolk. The two species sometimes mix and it is easy to confuse young grey seals with common seals. The former has a 'horse-shaped' head with parallel nostrils; the latter has a 'cat-shaped' head with nostrils in a V.

The breeding season of the grey seal runs from September to December; most pups at the Farnes are born in November. The cow seals come ashore and gather with other cows. They give birth a day later and thereafter feed the pup every five to six hours. Where the cows have pupped near the beach, they wait in the water and come ashore at feeding time, finding their pups first by recognising their position on the shore, then by voice and scent. In crowded colonies, the cows move on to ground above the shore and stay with their pups until they are weaned at sixteen to twenty-one days. The pups then shed their white coats and set out to sea by themselves.

Common seals breed in the summer, in June and July, but they are less tied to the land than the grey seal. The pups are born on sandbanks or rocks exposed at low tide, and they have to swim when the tide returns; occasionally a pup is born in the water. They are suckled on land or in the water. When tired they are carried on their mothers' backs and they are pushed under when danger threatens.

Both common and grey seals are generalist feeders, eating whatever is readily available. This is mainly fish, together with some crustaceans and molluscs. The diet of fish has brought them into conflict with fishermen, especially salmon netsmen. The seals steal fish from the nets, damage many more fish than they eat and also rip the nets. Furthermore, seals are hosts to the adult stage of the parasitic codworm. One of the worm's larval stages lives in cod and related fish, and infected fish have a reduced market value.

Until the early years of this century, the numbers of seals were kept low because their pelts and oil were sought by coastal dwellers. Such subsistence hunting has dwindled, and numbers especially of the grey seal, which was given legal protection in 1914, have increased considerably. Grey seals doubled in number between 1960 and 1970 on the Farne Islands, for instance, and the conflict with fishermen intensified.

Both species of seals can be very troublesome around fishing nets, and the Conservation of Seals Act 1970 allows seals to be shot to protect nets. More controversial has been the large-scale killing of pups and adults which was aimed at reducing the size of the population to protect the entire fishing industry. However, there is not yet any evidence that lowering the number of seals would raise the fishermen's catch and there is a good case for arguing that dwindling fish stocks are best protected by reducing the number of fishermen.

The National Trust has been faced with a very different problem at the Farne Islands, which illustrates the complexity of conservation problems. The Farne Islands have been famous for their sea-birds since the days of St Cuthbert, but their presence on the islands has been threatened by the increasing numbers of grey seals.

As the beaches became crowded, the seals moved on to the tops of two large islands, Brownsman and Staple, to bear their pups. Their movements tore up the plant cover and reduced the surface to a sea of mud. Not only was there severe soil erosion from the bare surface but the colony of puffins had its nest burrows destroyed. They had to dig new ones, which further increased erosion and, without a vegetation cover, rain soaked into the burrows and killed the nestlings. In places the soil disappeared at a rate of 2cm ($\frac{3}{4}$in) a year and bedrock was beginning to show.

A grey seal cow with her pup. Although flourishing in northern waters off Britain, grey seal numbers are declining in some areas, such as the south west.

Public opinion stopped an attempt to solve the problem by reducing the seal population, so, although a few seals are shot annually, since 1977 the National Trust has run an alternative 'disturbance plan'. The mere presence of wardens now keeps the pregnant cows away from the worst affected islands and they go to other islands to bear their pups. Meanwhile, the islands have been replanted with sea campion and grass, but this management plan is expensive and, if the islands are to continue to be shared by birds and seals, must continue for ever.

A kingfisher carrying prey to its nest.

PART TWO
Birds

7 *Birds of the Countryside*

RICHARD FITTER

Of all Britain's wildlife, it is the birds which are most readily seen and most easily identified. Not surprisingly birdwatching is one of Britain's most popular outdoor hobbies, with hundreds of thousands of people belonging to the Royal Society for the Protection of Birds and other bird clubs and societies. Not all these members are expert birdwatchers and many of the million plus members of the National Trust are also keen birdwatchers, adding up to a large part of the public with some sort of interest in birds.

Undoubtedly Britain's most popular bird is the robin. In fact, some years ago it was chosen by an overwhelming majority as our 'national' bird, and even those with little or no interest in birds would know a robin. The robin is also typical of one of Britain's most important habitats, and one that is well represented on National Trust properties – woodlands.

WOODLAND

Genuinely old woodland, which has been called wildwood, with continuous tree cover since the Middle Ages or even earlier, is now scarce in the British countryside, and becoming scarcer. The 1,000 acres of Hatfield Forest, a National Trust property in north-west Essex, are generally held to be as near as you can get nowadays in southern England to what woodland looked like in the Middle Ages. Nearly all our woods are either secondary (meaning that they have been either clear-felled and allowed to grow up again), have actually developed from cultivated land, or are mere plantations, mostly of alien conifers and of little wildlife interest. Nowadays there is not much difference between the birds of surviving patches of wildwood and those of woodland of more recent origin. What is more, the same group of woodland birds has, during the past thousand years or so, come out of the woods and colonised not only the extensive scrublands that are woods in the making after grazing or cultivation has been abandoned, but also hedges, which are linear strips of woodland and scrub; parks, which are often woodland with widely

Fieldfares (right) *and redwings. Both species are winter visitors to Britain, but small numbers have colonised Scotland.*

The dunnock is the classic little brown bird seen flitting among the undergrowth in woodland and gardens.

scattered trees and little or no scrub understory; and even large gardens, which are a splendid mixture of trees, shrubs, grass and bare soil that maximises the available habitat and is greatly attractive to many woodland birds, not least the blackbird, robin, song thrush, dunnock, wren and chaffinch. All these inhabit woods, hedges and larger gardens, and parks too where there are shrubs. And all these habitats are found in abundance on National Trust properties. Very large numbers of people every year visit gardens, parks and amenity woodlands of the Trust, and so have a chance of seeing a wide range of woodland birds.

In addition to the ubiquitous starling and house sparrow, about a dozen species of birds are likely to be seen in most fairly large gardens and they will thrive also in hedged farmland and any woodland that is not a close-packed plantation. On a walk in the well-wooded grounds of almost any of the National Trust stately homes or gardens such as Stourhead, Lyme Park, Ickworth, Sheffield Park, Nymans or Hidcote, most of the commoner woodland birds can be seen, together with several less frequently encountered. With the robin we associate also the sparrow-like dunnock (or hedge-sparrow) and the wren. All three, but especially the wren, sing almost throughout the year, and all three like the mixture of trees, shrubs and open ground that is found in woods, along hedges and in gardens. The robin has a pleasant thin warbling song, the dunnock a rather monotonous flat little ditty, and the wren a vigorous clear warbling song, surprisingly loud for the second smallest British bird.

After the robin, the species of bird most likely to be seen in woods, shrubberies and gardens is the blackbird. Although distinctive as an adult, the young (like those of robins) are very similar to thrushes; in fact, robins, thrushes and blackbirds are all closely related. Two species of thrush breed commonly in Britain: the song thrush, which is almost as widespread as the blackbird and likely to be encountered wherever there are blackbirds, and the larger mistle thrush. The latter is also known as the 'storm cock' from its habit of singing from the tops of tall trees, even in March gales. In winter two other thrushes migrate to Britain from Scandinavia – the redwing and fieldfare – and can usually be seen in company with blackbirds, song thrushes and mistle thrushes feeding on berries in hedgerows, or looking for worms and insects on playing fields and open grasslands. In recent years both the redwing and fieldfare have started breeding in this country in small numbers.

The tits are among the most popular birds – being frequent visitors to bird tables, and readily nesting in artificial nest boxes. They are also typical woodland birds. The most familiar tits are the sparrow-sized great tit and the much smaller blue tit. The great tit is one of the handsomer of our smaller song birds, with its bold black and white head pattern, lemon-yellow breast with a black bib, and blue-green mantle and wings. It is also one of the most, if not the most, versatile vocal performers. There are two terms used to describe birdsong: 'call' is the term normally used for alarm, flight contact and similar behaviour, whereas 'song' usually refers to the more elaborate sounds used during courtship. The great tit's usual calls are a loud 'pink' – much too like the chaffinch's for

Left *Typical English woodland.*

Right *Robins are highly territorial and the males will fight and drive off intruders.*

Left *Flocks of redwings feeding on hedgerow berries are a common sight in winter. Holly berries are among the last to be eaten, after the more favoured species such as hips and haws have disappeared.*

Right *The great tit is one of the most familiar visitors to gardens, and will readily nest in boxes.*

the comfort of even experienced birdwatchers – and clear, ringing 'teacher, teacher'. Its 'song' (most tits do not have a warbling song like thrushes and warblers) is the so-called 'saw-sharpening' note. The great tit also has such a wide range of other calls that almost any strange call heard in woodland is more likely to be a great tit than anything else. The blue tit has a smaller range of notes, its song being a gay little trill. Smaller still is the coal tit, which is like a miniature great tit, but with browner plumage and a conspicuous white nape. It prefers coniferous woodland, or at least mixed woodland with some conifers in it (as most large gardens also have), and its calls are also a miniature of the great tit's, especially its frequent contact note, 'teacher, teacher'. Similar again, but without the white nape, are the marsh and willow tits; their separation is difficult. The fifth woodland tit, the long-tailed, is usually seen only in the larger gardens. It has no song, just a spluttered call, often uttered as a flock flies, follow-my-leader, from bush to bush. It is well known for its unique oval nest, composed of thousands of feathers and decorated outside with lichens, held together with spiders' webs. In Scotland a sixth tit, the crested tit, easily identified by its crest, inhabits pinewoods and coniferous plantations from Speyside northwards to the south of Sutherland.

In late summer and autumn tits band together and roam through the woods and hedgerows often accompanied by such other woodland birds as the nuthatch, tree creeper, willow warbler and chiffchaff. The nuthatch and tree creeper can be seen running up tree

A coal tit taking off from a bird table. Feeding birds with scraps of bacon fat, peanuts, seeds and household scraps helps many species to survive the winter.

trunks or along large branches; the nuthatch also runs down tree trunks, the only British bird able to do this. They are easily separated by the nuthatch's contrasting plumage of blue-grey and pale chestnut, and short straight bill and the tree creeper's dull brown plumage and long curved bill. The nuthatch has a variety of loud ringing calls, especially in spring, but the tree creeper's song and call are both very high-pitched.

The final group of birds that have colonised our gardens from the woods and scrublands are the finches. The chaffinch is among the most widespread of British birds. Its gay little trilling song resounds wherever there are trees or tall bushes from late February until early July, and its 'pink, pink' note – it is usually a double one, not a single one like the great tit's – can be heard at all times of year. In winter, when chaffinch flocks feed on the fallen beech mast, they are often joined by the brambling, a winter visitor from Scandinavia, easily told by its white rump as it flies away. Greenfinches are birds of scrubland rather than woodland, but are frequently found in large gardens. The bullfinch, which also shows a white rump as it flies away, is a bird of woodland edges that is not uncommon in large gardens and orchards, and the redpoll, which frequents birches and alders in winter, has recently colonised many gardens around London. Along with the redpoll on the birches and alders, you can often see the siskin, which breeds in Scottish, and increasingly also in English, pinewoods. The crossbill also breeds in Scottish and other pinewoods, and sometimes crosses the North Sea from Scandinavia when its food supply there has failed, turning up all over Britain in late June or July. The hawfinch, our largest finch, rarely strays out of the woods into gardens.

There are only two species of sparrow in Britain: the house sparrow, all too common in gardens, and not in any way a woodland bird, and the tree sparrow, a bird of trees, such as those along rivers, rather than of woods, and not at all of gardens.

Every spring a whole new cohort of birds arrives in our woodlands: the migrants which have been wintering abroad, mostly in Africa. Most of these also inhabit scrub, especially if it contains scattered trees, and some also frequent the larger gardens, such as those around many National Trust properties, especially if, as at Cliveden, they are in a woodland setting. The warblers fill the woodlands with the sounds of spring – though they are often difficult to see. By far the most common woodland warbler is the willow warbler, distinguishable from its close relative the chiffchaff easily by its song, but only with difficulty by sight. No two closely related birds have such distinct songs as the willow warbler, with its liquid series of rather wistful descending notes, and the chiffchaff with its monotonous 'chiff-chaff' song. The willow warbler may inhabit scrub, but the chiffchaff demands at least a few trees in its territory from which to sing. The wood warbler has two songs, a long, quivering trill and a repeated plaintive disyllabic note, so different that the birdwatching novice could well be forgiven for thinking two different species were singing. These three are all collectively known as 'leaf warblers'. The goldcrest, our smallest bird, is also a warbler, looking like a tiny leaf warbler and demands a conifer, which may be a yew, in any wood in which it nests. Its slightly larger relative, the firecrest is rare, but slowly spreading.

The blackcap is the easiest warbler to identify, with the male's black and the female's brown cap only confusable with the marsh or willow tits. Its close relative, the garden warbler, has fewer distinctive markings than almost any other British song bird, being plain brown all over.

With its rich warbling, trilling and musical song the nightingale is related to the thrushes and the robin, not the warblers. Dense shrubberies, woodlands with thick undergrowth and scrubby heathlands – such as the National Trust's Dunwich Heath – are all favoured habitats of the nightingale. Despite popular belief, not every nocturnal songster is a nightingale; in fact almost invariably those claimed in towns and cities turn out to be blackbirds or song thrushes.

Like all other game birds, pheasant chicks leave the nest soon after hatching and follow the mother. To escape detection when small, they rely on 'freezing' and their camouflage.

On a walk around a woodland, only a few other birds are likely to be conspicuous: the garish and noisy jay, which is particularly obvious in late autumn when acorns are abundant; or the tawny owl, which is less likely to be seen, though if you notice jays, blackbirds and small birds making a lot of noise and repeatedly diving at a particular spot it is likely to be an owl being mobbed.

The pheasant is possibly the most spectacular of all woodland birds – but difficult to take seriously as a wild bird since thousands are released by gamekeepers every year – it can also provide a considerable shock to someone quietly walking in the woods, when it explodes from the undergrowth with noisily whirring wings.

The redstart cock has one of the most gorgeous plumages of any of our song birds, with its white forehead, black throat and grey mantle offsetting its fiery red tail and underparts. Its song, however, is undistinguished. The black redstart, with an equally disappointing song, is a bird of towns, industrial sites, and only occasionally sea cliffs. The pied flycatcher is the only small black and white bird likely to be seen in woodland and, like its dull grey-brown relative the spotted flycatcher, is easily told by its habit of perching on a prominent twig and sallying forth and back again after flying insects. The pied rarely enters small gardens, but the more widespread and abundant spotted flycatcher may actually nest on a house wall.

There remain two important groups of woodland birds, less often seen in small or medium-sized gardens than most of those above: the woodpeckers and the pigeons or doves. We have four native woodpeckers, but one of them, the small brown wryneck, is

Opposite *A female greater spotted woodpecker feeding her well-grown young.*

Below *A cock redstart bringing a beakful of insects including an earwig to the nest.*

now very rare. Easily the most handsome is the gaudy green woodpecker, which many people, on first sighting it, assume must have escaped from a zoo. The contrast of its red top-knot with its green and yellow plumage, and its loud ringing cry both seem rather exotic in the quiet British countryside. In the parklands of the National Trust properties, dotted with mature, isolated trees and grazed by cattle, the green woodpecker is not an uncommon sight, feeding on the ground for ants in late summer and autumn. Our two other woodpeckers are the greater and lesser spotted; the latter is much less common and hardly bigger than a sparrow. In recent years the numbers of the greater spotted woodpecker have increased particularly in suburban areas, where it is often very destructive to nest boxes and even preys on baby tits. Of our five native pigeons or doves, three frequent woodlands: the ubiquitous and pestilential wood pigeon, the one British bird to do serious economic harm; the stock dove, usually seen in old parkland, where it nests in holes in the ancient trees; and the turtle dove, which is a scrub rather than woodland bird, and takes its name from the soothing 'turr-turr' note that passes for its song. Our two other doves are the recently arrived collared dove, usually found near places where chickens or other domestic birds are fed, and the rock dove of our northern and western cliffs, the bird which gave rise to all our domestic pigeons and the feral pigeons of our towns.

HEATHS AND MOORS

For a walk in more open country, we can go either through the scrublands to heaths and moors or straight into the hedged countryside. Whichever we do, we shall still find many of the woodland birds, but will also encounter many new ones which are not likely to be seen in a wood. Most heathland offers a range of habitat, from scrub with small trees to open heathery or grassy areas. In the south most heathlands are such a mosaic, but in the north there are extensive stretches of quite treeless grass and heather moors.

One of the most striking birds, characteristic of the southern heathlands, is the nightjar, rarely seen by day, but more easily heard making its curious monotonous churring note at night, in the New Forest for instance; its numbers are, alas, decreasing. The stonechat population has also decreased, but it can still be found on many gorse-covered cliffs in the west, including those of the National Trust in Cornwall. The whinchat, a summer visitor, is perhaps more frequently seen on the lower moors than on the heaths at sea-level round the coast. Wheatears, one of the earliest migrants to appear (often by mid-March in the south), frequent both heaths and moors. The woodlark, one of our choicest songsters, although its numbers greatly decreased after the hard winter of 1963, seems to be regaining some ground. You are also likely to hear the tree pipit, which is easily recognised by its song-flight, from a tree and back to a tree, uttering a short song ending in 'see-er see-er see-er'. But to tell it on the ground from the much commoner meadow pipit is difficult. Meadow pipits do breed on heaths and downs, but are the characteristic bird of the moors, along with the skylark. Another bird which frequents both heaths and moors is the cuckoo, whose favourite victim is the meadow pipit.

Where there are bushes on the heathlands, you will find linnets, whose moorland counterpart is the twite. The nearly extinct red-backed shrike, whose long-term decline in Britain is not fully understood, is now a rarity of bushy heathland, but until the 1950s it was not an uncommon sight to see this bird – also known as the butcher bird – impaling its prey, such as grasshoppers or even small lizards, on thorns or barbed wire. The Dartford warbler is also rarely found on bushy heathland now. It was very nearly exterminated by the 1963 winter, but has since recovered. It prefers thick gorse, but can make do with very long heather.

The two typical birds of the open moorland are the meadow pipit, as already mentioned, and the red grouse. Unlike the black grouse, which lives along the woodland fringes of the moors, and the capercaillie, confined to the pinewoods of the Scottish Highlands, the red grouse inhabits treeless heather moors. On the high tops, along with the dotterel and the occasional snow bunting, and again only in the Highlands, the red grouse is replaced by the ptarmigan, which turns almost white in winter. They are both preyed on by the golden eagle.

The ring ouzel, the white-breasted blackbird of the moors, is too small to be the victim of an eagle, though it may well fall to the swift attack of the smallest of our falcons, the merlin, but only to the female, for the male is scarcely larger than a ring ouzel itself. Two other birds of prey, the buzzard and peregrine falcon, along with the raven, our largest crow, are as frequent along cliffed coastlines as in the hills. The raven's distinctive relative the hooded crow is, however, commoner on the Highland moors than along the cliffs, while the jackdaw is a bird of cliffs everywhere, and the now very local chough is found mainly on sea cliffs and only in a very few places in the hills inland, all in Wales.

We cannot leave the moorlands without mentioning their importance for breeding waders: lapwing, curlew, golden plover are widespread, whimbrel is very local in Scotland, and dunlin is found on some of the higher tops.

The National Trust owns, or has covenants over, extensive tracts of moorland, including nearly 10,000 acres in the Brecon Beacons, several parts of the Yorkshire moors, the Lake District, the Derbyshire uplands and heathlands in Surrey and Hampshire.

FARMLAND

Farmland is one of the National Trust's most extensive holdings, most of which is farmed by tenants, but usually with special concern for preserving the traditional features of the landscape. With increasing 'modernisation' and the aggregation of many farms, the traditional landscapes with hedgerows and copses are becoming an increasingly important feature of properties owned and managed by the National Trust.

Just as the woodland birds spill out into the heaths and moors, so they do into farmland, especially in a hedged countryside, with scattered trees or clumps or copses in the hedgerows. Blackbirds, song thrushes, robins, hedge-sparrows and wrens are all well-

Left *Rooks start to nest when the trees are still bare. The disappearance of elm trees in much of Britain has displaced many rookeries.*

Right *House martins gathering mud to build their nests under the eaves of houses. In a few places house martins still nest on cliffs and rock faces – their only sites before buildings.*

known hedgerow birds, and so (among the birds we have already met) are chaffinches and greenfinches. Goldfinches are more birds of open country, feeding in late summer and autumn on thistle-heads, but in spring their favourite nesting place is an orchard tree. Yellowhammers do not nest very high up in hedges, but there are nearly always hedges in the countryside where they are found. Their close relative, the cirl bunting, is now quite localised and rare, perhaps because it had a liking for elm trees, which have all but disappeared from the landscape in the aftermath of Dutch elm disease.

The whitethroat, a warbler, is also commonly associated with hedges, from the top of which it rises to utter its scratchy little song. The lesser whitethroat, on the other hand, though it usually nests in a shrub, seems to need small trees and more mature hedges, and utters its monotonous rattle from deep within them.

Crows of various kinds and starlings are usually abundant in farmland, where the grass and arable fields provide them with feeding places and the trees with places to nest. Rookeries are familiar sights, though less so than a few years ago, again because of the felling of elms which have died through Dutch elm disease, and the single nests of carrion crows are still common in areas where game-keepering is not vigorous. The distinctive magpie, with its conspicuous domed nest, is another victim of the game preserver, and was at one time virtually extirpated from the county of Norfolk by gamekeepers. But it is now a highly successful colonist, moving into city parks and gardens. Jackdaws nest in old buildings and, like stock doves and tawny owls, in the large old trees in parks. In winter they often feed and fly along with the rook flocks. Wood pigeons often nest in hedges and collared doves frequent the neighbourhood of farms, nesting in trees and shrubs near by.

Out in the open fields many birds we have already mentioned feed freely, mixed flocks of finches and yellowhammers, as well as starlings, rooks, jackdaws, wood pigeons, lapwings and in some areas golden plovers. But there are also birds that breed as well as feed out in the open, most notably the two species of partridge. The common or grey partridge is now quite scarce in many areas, victim of the excessive use of farm chemicals – they kill off all the insects on which the partridge chicks feed. So the chances are that a partridge covey will consist of the introduced red-legged partridge, often called the Frenchman. Quail are still very uncommon and local birds, mainly in chalk districts, and the corncrake has retreated to the fastnesses of the western Highlands, the Outer Hebrides and Ireland. The stone curlew is a local bird, and largely restricted to the grassy downs of the southern chalk and East Anglia. Two small birds that can breed in a totally hedgeless countryside are the skylark, which we have already met on the moors and which is very common on

Willow warblers nest close to the ground and, in common with other warblers, the young are fed almost exclusively on small insects and spiders.

downland, and the corn bunting, our largest bunting, whose jangly little song can be uttered from a clod of earth or a telegraph wire. The yellow wagtail too will nest right in the middle of a field, but the 'improvement' of meadows has led to a decline in this species in many areas since the meadows lack the rich fauna of insects, and in any case are often cut for silage. But the pied wagtail is usually found nearer the farmhouse itself, nesting in some cranny of the building, along with swallows, house martins, swifts, house sparrows, starlings and spotted flycatchers, all with their specialised niche. The house martins are the ones whose nests are most conspicuous, plastered against the house wall under the eaves; many people think that these nests belong to swallows, but swallows usually nest inside a building, on a beam.

No account of farmland birds can be complete without mention of the predators, which are dealt with elsewhere: kestrel, hovering over the fields; sparrowhawk, dashing along the hedgerows; barn owl, ghostlike in the gloaming; and the little owl perched, bobbing inquisitively, on a telegraph pole.

FRESH WATER

Finally, some mention must be made of the freshwater birds which are neither true waterfowl nor waders, and so dealt with elsewhere. The heron stands gaunt and watchful by

Left *Around many of the larger lakes in National Trust properties the grey heron is a familiar sight, standing motionless ready to snatch a passing fish.*

Right *The nest of the reed warbler is built around reed stems. Reed warblers are one of the species often parasitised by cuckoos.*

the rivers, lakes and estuaries, usually nesting in trees near by, but the bittern is a rare bird of extensive reedbeds mainly in East Anglia, for instance at Horsey, Norfolk. The water rail rarely comes out of its marshy habitat to be seen, except when driven by frost and ice. The kingfisher is another sufferer from hard winters, and takes two or three years to recover its numbers after each such winter, as 1981–2. Swallows, martins and swifts all feed commonly over fresh water, but the sand martin frequently breeds near by, either in a gravel pit or in the sandy banks of a stream. The reed bunting is a common bird of marshes and watersides but is found increasingly on ordinary farmland. The bearded tit is a reedbed denizen, once confined to East Anglia but now spreading to a few other extensive swamps, mainly round the coast. As with the bittern, Horsey is a good place to see it. It is not a true tit but is allied to the babblers of the tropics.

Six species of warbler are especially associated with fresh water. The reed warbler is rarely found away from reeds, whether extensive reedbeds, or strips along a watercourse. The sedge warbler, distinguishable by its pale eyestripe, is more catholic in its choice and will nest in almost any waterside shrub or thick vegetation. The rare marsh warbler, very hard to distinguish from the reed warbler, except by its highly mimetic song, nests in a few osier beds in the West Country. The grasshopper warbler is the most secretive, being rarely seen, but very audible, thanks to its song which sounds like an angler's reel. It may also nest in drier places than the others, for instance on heathland or the site of a felled woodland. Two recent invaders of our watersides and marshes are Savi's warbler, a relative of the grasshopper, and Cetti's warbler, more like a reed warbler, and one of only two warblers – the other is the Dartford – that regularly winters here. Cetti's warbler is most easily recognised by its staccato bursts of loud singing from the depths of a thicket, from which it does not like to emerge.

Finally, there is one bird confined to the seaside that does not rank as a sea-bird. The rock pipit, like a greyer meadow pipit, is frequently seen on rocky shores all round the coast, such as those owned by the National Trust in many parts of Cornwall.

8 *Waterfowl*

Dr MYRFYN OWEN

Waterfowl are swans, geese and ducks, which are classified by scientists in the family *Anatidae*. They are familiar birds, characteristically associated with water and wetland habitats. The commonest is the mallard, but tufted duck, pochard, teal, shoveler, shelduck, mute swan and Canada geese are all regularly seen on many National Trust properties. These birds range in size from the diminutive teal, weighing about a pound to one of the heaviest flying birds – the native male mute swan – which reaches 14kg (30lb) or more in weight. Wildfowl are strong, fast fliers (the fastest flying bird is probably the

Left *Mute swans fighting. The immature swans are greyish brown, gradually moulting into the pure white adult plumage.*

Right *The greylag goose is the ancestor of the farmyard goose.*

The mute swan is one of the heaviest flying birds and needs a long take-off path.

eider duck) and most British species are migratory, breeding in the north temperate or Arctic regions of Europe and flying to winter in the milder regions of coastal north-west Europe and the British Isles. Several other birds, such as divers and grebes, are superficially similar.

SWANS

Swans are large and (with the exception of two southern hemisphere species) have all-white feathers. They are associated with many legends and have long been regarded with respect, even reverence, by man. The three species wintering in Britain are all protected, the most familiar being the mis-named mute swan, which not only makes a musical whistling sound with its wings while flying, but also utters a variety of grunts and hisses. The mute swan is a resident breeder in Britain, nesting in inland lakes and rivers and moving only very short distances in winter. Like all swans, the mute is highly aggressive

Left *When migrating greylag geese usually fly in V-shaped skeins, but when moving between feeding grounds the flocks are less organised.*

Right *Pink-footed geese grazing in Scotland.*

and territorial during the breeding season, using its bright white plumage for advertisement and chasing away any intruder that ventures near its large nest.

Another feature of swans is the life-long pair-bond, formed when the birds are three or four years old. The cygnets are brownish-grey for the first year and then gradually assume their mature dress. The aggression of breeding pairs is largely lost in winter, when flocks are formed on lakes and estuaries rich in submerged aquatic plants on whose leaves the swans almost exclusively feed. The largest flock is that resident on the Chesil Fleet in Dorset (part of which is owned by the National Trust), with numbers sometimes exceeding a thousand birds, but smaller concentrations are often found in populated areas where the swans are tame because of their protected status, and take advantage of bread and titbits supplied by the public. Similarly, they are often to be found on the larger lakes in the grounds of stately homes such as Stourhead and Tatton Park. Recently there has been considerable concern over local declines caused by damage from discarded fishing tackle. The main threat to the mute swan comes from discarded lead weights; many other waterfowl suffer slow and painful deaths from swallowing hooks or getting entangled in nylon line. The general public can help here by removing any hooks, line or weights discarded on river banks or lake sides and disposing of them safely.

The other two species, the whooper from Iceland and the smaller Bewick's swan from Siberia, migrate here in October and November. Both are vegetarians, the whooper swan staying largely on the lochs and estuaries of Scotland and Ireland and the Bewick's on the floodlands and freshwater marshes of southern England.

GEESE

Geese are closely related to swans and share many aspects of their behaviour such as life-long monogamy and the maintenance of the family through the whole of the first winter. Geese are also highly gregarious outside the breeding season, migrating together in the characteristic V-shaped skeins and wintering in large concentrations. Geese are separated into 'grey' geese, which have brownish or greyish plumage with orange or pink bills and feet, and 'black' geese, which have black soft parts and a substantial amount of black feathering.

The only native British goose is the greylag, now driven from the marshes and fens of England to its only refuge in the Outer Hebrides. Flocks have, however, been reintroduced, largely by wildfowlers, in the south and there is now a feral population of several thousand in Britain. The majority of wintering greylags and all pink-footed geese come here via Iceland to spend the winter predominantly in Scotland. White-fronted geese from Greenland also stay in Scotland and Ireland, while small numbers from the vast Siberian breeding flocks, cross the North Sea to terminate their winter's stay in southern England, on estuarine saltmarshes and river floods.

Of the 'black' geese, the barnacle goose breeds in Greenland and Spitzbergen and winters in western Scotland and Ireland and on the saltmarshes of the Solway Firth. The brent goose, the smallest of our species, is predominantly black and is largely restricted to coastal marshes of eastern and southern England, from Lincolnshire to Devon. The brent visited Britain in tens of thousands to feed on the eelgrass – *Zostera* – on the muddy east-coast estuaries until a wasting disease affected that plant in the 1930s and brent numbers declined drastically. A recent revival followed protection from shooting throughout its most important haunts, and the brent became so numerous that its native saltmarshes (which are continually being eroded by reclamation) could no longer contain it. The result was that flocks overflowed to graze pastures and winter wheat, and the species incurred the wrath of farmers. The protection of the remaining saltmarshes is crucial to the well-being of the brent and the National Trust holdings in north Norfolk, particularly around Scolt Head Island, where several thousand geese winter, are making a substantial contribution.

The Canada goose, introduced into Britain as an ornamental bird some three centuries ago, is now the most widespread and numerous goose of inland waters. The introductions were made on to the lakes of large estates and stately homes in the Midlands and south of England and the largest flocks (some are over 300-strong) are still found there. The geese were allowed to breed freely on islands in the lakes and their grazing on surrounding lawns and pastures was tolerated or even encouraged. The flocks of rather tame Canada geese have become a popular feature of homes open to the public such as Clumber Park in Nottinghamshire, Tatton Park in Cheshire and Petworth Park, Sussex, all owned by the National Trust. While the increase in Canada geese is welcomed by many, there is increasing concern about their impact on agriculture and there have been calls for their control.

DUCKS

Intermediate between the geese and ducks, the shelduck is a brightly coloured and common bird of our estuaries, breeding in holes in sand dunes and banks and wintering in

Top left *pochard;* top right *shelduck;* above left *golden eye displaying;* above right *drake wigeon.*

large numbers on our major intertidal stretches. Here it feeds on the tiny water snails and other invertebrates.

Ducks are the most diverse group of wildfowl and up to thirty species are regularly seen in Britain. One of their most striking features is the colourful plumage of the drakes in contrast to the drab females. With very few exceptions only the female incubates and cares for the young; the males gather in secluded waters to moult as soon as the duck begins to incubate.

Ducks can be divided broadly into dabblers and divers depending on their feeding methods. Most common of the dabblers or surface-feeders is the mallard, a ubiquitous breeder in inland Britain, living in lakes, marshes, farmland and town parks and to be seen on virtually every National Trust property with a lake or pond. The large British breed-

The great crested grebe builds a raft-like nest which floats. The young leave the nest soon after hatching and are carried on the backs of their parents while small.

ing population is supplemented in midwinter by immigrants from the near continent, Scandinavia and the USSR. The teal – a small species with the drakes having bright brown and green heads – also breeds in Britain in small numbers but the vast majority are winter visitors from their breeding grounds in north-west Europe.

The wigeon's musical whistle is a familiar estuarine cry; it is our second most numerous wintering species. It breeds in small numbers in Scotland; but most of the wintering flocks travel 2,000–3,000 miles from their arctic Siberian breeding area. The wigeon is a grazer, feeding on saltings and pastures in large and densely packed flocks. The closely related gadwall, though occurring in small numbers as a wintering bird, has been introduced to parts of England in recent years and is increasing as a resident breeding bird. Among the other dabbling ducks likely to be seen in moorland habitats are the shoveler, with its bright brown, white and green plumage and spathulate bill; the elegant pintail, a visitor to our large estuaries; and the garganey – the only duck which is a summer visitor (though in declining numbers).

The diving ducks include the vegetarian pochard, which is chiefly found in inland waters, especially gravel pits, reservoirs and ornamental lakes in the lowlands. The most familiar inland species is the tufted duck – the males bright in their black and white winter plumage. Chiefly an animal feeder, this species breeds here in large numbers and has been a major beneficiary of the recent great increase in the area of gravel pits and reservoirs. Some of our wintering tufted ducks come from the east, while part of our breeding stock move to winter in Ireland and France. The scaup and goldeneye are mainly found on the coast but the latter occurs in small numbers on inland waters.

The so-called sea-ducks are truly marine in winter and include eiders, scoters and the long-tailed duck. The eider is the second most numerous breeding duck in Britain, the bulk of the birds nesting in northern Scotland. Eiders never wander far from the sea, so the nests are on islands or peninsulas, protected from predators by their isolation and the highly camouflaged plumage of the female. The black and white plumage of the males is not fully developed until they are two years old. British eiders are resident all the year round, wintering in large concentrations chiefly off the Scottish coast. The largest colony

A tufted duck taking off from a frozen lake. They are the most common of the diving ducks and likely to be seen on almost all the larger lakes in National Trust properties.

in England, numbering more than 1,000 pairs, is on National Trust land on the Farne Islands. Many winter locally but some travel northwards to the shallows of the outer Tay estuary. Like other sea-ducks, the eider depends mainly on molluscs for its food, and in particular the edible mussel, which it tears off the rocks with its strong bill and swallows whole. St Cuthbert was a hermit who lived on the Inner Farnes, and he is alleged to have tamed the eiders – which even to this day sit tight upon their eggs and can be stroked by an intruder!

The other sea-ducks are much less familiar, since they are not commonly seen from land. They gather in rafts over mussel beds offshore, mostly around northern Scotland and Wales. Only a few common scoters breed here, velvet scoters come from Scandinavia and the USSR, while the long-tailed duck breeds in Arctic areas of northern Europe.

The most highly adapted to an aquatic life are the fish-eating sawbills. The goosander is a resident breeder in inland Scotland and northern England, while the red-breasted merganser breeds inland and winters on the coast, again in the northern part of the country. Both come into conflict with fishing interests since a major part of their summer diet consists of salmon and trout fry. The tiny black and white smew is an uncommon visitor from north-west Europe, most being found in small numbers on the reservoirs of the London area.

OTHER WATERFOWL

In addition to the true wildfowl, a number of other unrelated, but superficially similar birds can often be seen on open waters and other wetlands.

The mainly marine cormorant often comes inland, and at one time was a breeding bird in the marshlands of East Anglia. On the larger lakes and rivers on National Trust properties both the great crested and little grebes are to be found as breeding birds, while other grebes turn up from time to time mainly as winter visitors. Like the grebes, divers are diving, mainly fish-eating birds. The red-throated diver breeds in northern Ireland and Scotland, the black-throated in Scotland and both species, together with the great northern diver, are to be found at sea and sometimes on larger stretches of inland waters (such as reservoirs) in winter.

Although like wildfowl, in that they are always associated with water, the moorhen and coot are both rails. Coots are often particularly abundant on reservoirs in winter, while almost any overgrown pond, however small, will have a pair or two of moorhens skulking in the reeds. Both species are predominantly black; the moorhen has a red front to the head, the coot has white.

HABITATS

The most characteristic feature of wildfowl is their dependence on the wetland habitats. Even those species such as geese and wigeon which gather their food on dry land rely on estuaries or lakes to provide a night-time roost safe from foxes. Most inland species depend on ponds and lakes and in the south of the country these are mainly artificial, resulting from reservoirs, gravel or clay-digging or from mining subsidence. Reservoirs for drinking water and canal headwaters especially in the nineteenth and early twentieth centuries provide a large area of suitable wildfowl habitat. Lakes are associated with a large number of estates and stately homes, as in many of the National Trust's holdings. Many were specially created and stocked with ornamental wildfowl which provided a focus for their colonisation by wild birds.

Our rivers and canals provide a home for relatively few species, but they become very important if in winter they overflow their banks and flood large areas. The most important inland site for wildfowl in Britain is just such an area, the Ouse Washes in East Anglia, which provide a reservoir for the floodwater from Bedfordshire and Cambridgeshire.

Inland wildfowl have fared well in the last two decades, although floodlands are constantly threatened with drainage as pressure builds up from farmers to squeeze every ounce of agricultural production from marginal land. Nearly all fenland in England has been lost and the largest remaining inland area of wetland, the Somerset Levels, has either been drained or is under threat.

Estuaries are very important wildfowl haunts, especially in severe weather, when not

Left *Wicken Fen, an important wildfowl habitat, was one of the first properties to be acquired by the National Trust.*

Right *Stourhead, Wiltshire, one of many Trust properties with extensive parkland and lakes which attract ducks and grebes.*

only do our inland birds move in search of open water but we also receive substantial numbers of refugees from the cold continent of Europe. Because of the continental origin of migrants, the estuaries on the east coast are vital.

Agricultural and industrial man continue to encroach on to our major estuaries, making the conservation of the remaining intertidal area even more crucial. Some species, like the shelduck, brent goose and pintail are largely dependent on this habitat and their future will depend on its preservation.

Being a major owner of wetlands, the National Trust has an important part to play. Inland it provides a large number of relatively undisturbed areas and many sites, such as Wicken Fen in Cambridgeshire and Horsey Mere in Norfolk, not only preserve important examples of habitat but also hold a wide variety of wildfowl, many in important numbers. The major contribution is, however, to be made on the coast. The importance of the marshes of north Norfolk for brent geese and estuarine ducks is undisputed. In other sites such as the Newtown estuary on the Isle of Wight, Brownsea Island in Poole Harbour, Dorset, and the Burry Inlet in south Wales (not National Trust), parts of these major complexes managed as reserves provide vital refuges from disturbance.

As permanent and sympathetic landowners, the National Trust can make a vital contribution to the conservation of Britain's dwindling wetlands and the wildfowl which depend on them for their survival.

9 *Waders*

JOHN GOODERS

The coastline of Britain is marked by a variety of scenery unsurpassed in Europe. In the north and west the sea has cut deep into the hills, creating spectacular cliffs broken here and there by picturesque fishing villages such as Clovelly in Devon. In the south this same sea action has eaten deep into the chalk giving rise to the white cliffs of Dover, the Seven Sisters, St Margaret's Bay and Beachy Head. In the east the sea has worked in another way depositing material eroded elsewhere to form islands and shingle banks such as Scolt Head and Blakeney Point.

All of these coastlines in their different ways attract visitors throughout the year and the National Trust is responsible for many of the most beautiful and significant areas. Many are important to wildlife and, in particular, support some of the most spectacular sea-bird colonies in Europe. Yet between these natural wonders there are areas of coastline that are largely ignored – lonely coastlines where the walker may stroll all day with only the birds and the occasional marshman for company. These are the estuaries.

Strange as it may seem estuaries are among the richest of all habitats, producing more biomass (i.e. life) than even the most fertile of corn fields. Twice each day the tide washes over the sand and mud banks bringing a wealth of planktonic life that is left behind as the water ebbs. These microscopic forms of life are food to a host of marine creatures that spend the period of low tide hidden in the ooze. These in turn are food to a host of birds – the waders.

Not surprisingly these birds have adapted to the feeding opportunities in a variety of ways. Some have remarkably long bills to probe deep into mud or sand for the large rag and lug worms. Others have chisel-like bills to hack open mussels and cockles. Some sift out the plankton along the tide edge itself, while still others pick food from among rocky clefts and fissures.

Because estuaries are so rich in life – there may be as many as 10,000 creatures in a cubic metre of mud – waders often congregate in huge flocks. Indeed they may gather at a particularly favoured feeding ground in such numbers as to make even the teeming sea-

A flock of dunlin in winter plumage. Dunlin are among the most common waders in Britain and flocks can be seen on many estuaries and mudflats during the winter.

bird colonies of such places as the Farne Islands seem bare in comparison. But while the Farnes attract visitors galore, the estuaries are largely neglected by all save the dedicated birdwatcher.

With a few exceptions waders are generally dull-coloured birds outside the breeding season. The lapwing is one of the most familiar waders as it circles and dives over the fields in spring, and it often outnumbers the rook on the damp fields of autumn and winter. Its 'pee-wit' call is responsible for one of its many country names and is almost as familiar as the call of the curlew, which introduced a well-known radio nature programme. Long-billed and long-legged the curlew is the archetypal wader. The avocet is a spectacular black and white wader that returned to breed in East Anglia after the Second World War following an absence of a hundred years, better known to many people as the symbol of the Royal Society for the Protection of Birds than through personal acquaintance. So lapwing, curlew and avocet we know, but there are thirty or more species that remain largely unseen and unknown by the non-bird enthusiast.

Walk the winter marshes out to Blakeney Point in Norfolk at low tide and it may seem a bleak, uninspiring sort of place. There will be odd groups of birds here and there, but usually much too far away to identify without the aid of a powerful telescope. As the tide rises the birds are concentrated into a smaller and smaller area of open mud until eventually they all take to flight in tight formation, heading for a safe high-tide roost among the saltings. Dunlin may be the most common species and flocks several thousand strong wheel in the air, exposing first their dark upper side then their pale under parts. The aerial evolutions are truly spectacular as each bird changes direction in complete synchronisation with its fellows. Finally all will settle at their roost to doze away the hours when the tide covers their feeding grounds.

Dunlin breed on upland moors throughout Britain and northern Europe. In summer they sport a black belly and in early autumn the adults can be distinguished from young birds of the year by the unmoulted remnants of this nuptial plumage. In winter they are grey and nondescript and to be identified by their longish, slightly downward-curved bills and ever-active feeding habits. A close view from well-sited hides such as those at Brownsea Island in Poole Harbour reveals just how hard these little birds have to work for their food. Probe, probe, prod, prod – it's non-stop action hour after hour.

Like the dunlin, most of the waders that frequent our estuaries breed elsewhere. Many come from Scandinavia and northern Russia, where the spring melt of ice and snow creates huge areas of marshy pools on a foundation of permanently frozen soil. Some, like the knot and sanderling, come from central northern Siberia or Greenland to spend the winter around our shores. Others, like wood and curlew sandpipers, breed in these northern fastnesses but only pass through Britain on their way south to the milder climes of Africa. These, along with the bar-tailed godwit, were called 'the globespanners' by the naturalist-hunter Abel Chapman, who knew the marshes around Lindisfarne in Northumberland long before the National Trust became involved in their protection.

Some waders breed with us, some winter, and some simply pass through on their way to and from their breeding grounds further north. Others, such as the oyster-catcher, are with us all year round. This large black and white bird with the huge, orange-red bill breeds right round our coasts and its piping call is a familiar sound on all but the busiest (in human terms) of shoreline beaches. Despite its name it is not really an oyster-catcher; mussel-catcher would be far more appropriate.

Oyster-catchers fall into two types – those that hammer their way into the hard shells of mussels and those that prise open the shell to extract the animal inside. These birds live

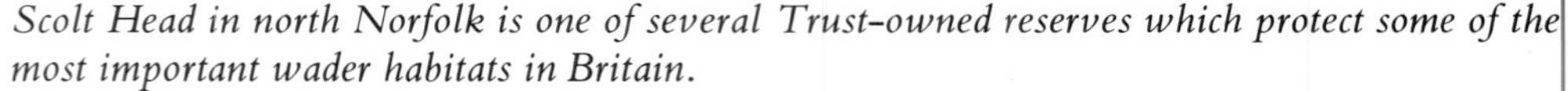

Scolt Head in north Norfolk is one of several Trust-owned reserves which protect some of the most important wader habitats in Britain.

Oyster-catchers are found around most of Britain's coastline, where they feed mainly on mussels.

side by side and it appears to be a matter of learning as to which type an individual becomes. In winter oyster-catchers gather in huge numbers at mussel-rich estuaries such as the Burry Inlet in south Wales. Fishermen here and at Morecambe Bay in Lancashire have protested that these birds are seriously reducing their livelihood, though most ornithologists would argue that there must be a surplus of mussels to attract the birds in the first place.

Never as obvious, but equally as common and widespread as the oyster-catcher, is the redshank. Dull brown in colour this medium-sized wader can be found on every British estuary and at many places inland as well. Its sharp yelping calls are a familiar sound to all who know the estuaries, and not for nothing is it called 'the sentinel of the marshes'. Its long red legs are the best clue to its identity, though there are other less common species that share such a feature. One of these is the ruff, though its legs are generally more orange than red.

Last century the ruff bred among the marshes of East Anglia at places like Wicken Fen when this unique fenland habitat was much more widespread than it is today. (Out of over 1½ million acres of fen, Wicken Fen's 730 acres are all that remain.) At that time the fensman took advantage of the male ruffs' communal dancing display to net them for the table. Up to a hundred or more males might gather at a traditional dancing ground, called a lek, to vie one with another for the right to mate with the females, the reeves, that came to visit. Each male was bedecked with brightly coloured ruffs and plumes that were shown off like an eighteenth-century dandy. Unfortunately a combination of fenland drainage, netting for the pot and the mania of the specimen collector finally eliminated the ruff as a British breeding bird. However, it has now returned once more and can be seen at its leks in various parts of the country including the Cley marshes in Norfolk. The adjacent Arnold's Marsh can be seen from the ornithologically famous East Bank, and here ruffs in all their finery can be seen throughout the summer.

Black-tailed godwits have also come back after suffering a similar fate to the ruff. At first their return to the Ouse Washes of Cambridgeshire was a closely guarded secret, but now they have spread elsewhere and can be seen at Cley, Salthouse, and other places along the north Norfolk coast. Though they have increased to over a hundred breeding pairs, this is as nothing compared with the numbers that pass through on their way to breeding grounds in Holland. Many of these birds, delightful in their rust-red nuptial plumage, gather in East Anglia before flying the short distance across the North Sea. Indeed their numbers were sometimes so great that May 15th is still known as 'Godwit Day' in parts of Suffolk and Norfolk. The famous wader grounds at Hickling Broad are a fine place to see these and hosts of other passage waders in both spring and autumn.

While the return of several waders to breed after years of absence can be attributed to a more understanding attitude toward wildlife in recent years, there have also been changes due to other factors. Decreasing summer temperatures have brought Scotland within range of a host of Scandinavian breeding birds including the wood sandpiper, Temminck's stint and the purple sandpiper. Meanwhile in the south we have been colonised by the little ringed plover, a bird that frequents the shingle banks of continental rivers, but which has found the edges of freshly created gravel pits a satisfactory substitute.

Recent history has not been all gains, however. In southern England the Kentish plover was eliminated in the 1950s and, despite rumours and the occasional confirmed breeding, it has yet to make a permanent return. A quite different wader, the stone curlew, is fast disappearing as an inhabitant of dry sandy wastelands. Once numerous on the Breckland of East Anglia and throughout the southern chalk, post-war ploughing of marginal land has destroyed huge areas of suitable habitat. So the stone curlew turned to arable fields and its nests were destroyed by agricultural machinery and the quick-growing strains of cereals. Now down to less than five hundred pairs, there are few areas left for this strangely crepuscular bird with the tin-whistle call.

Like the stone curlew, the woodcock is an aberrant wader that has forsaken the estuary and adapted to a quite different habitat. As its name implies it is a woodland bird feeding in damp marshy spots and nesting among the dead leaves of the forest floor. Its plumage so resembles dead leaves that it relies entirely on camouflage for concealment and its nest is remarkably difficult to find. Throughout the day it hides in the forest and flies out only under cover of darkness to feed. It is most easily located on fine summer evenings when the males perform a strange roding flight round and round their territories uttering shrill 'tsick' flight notes.

These then are the strange, one could even say abnormal, waders, waders that have adapted to a way of life that has taken them away from the lonely marshes. For really typical waders and the thrill of wild birds in their tens of thousands it is the coastal marshes and estuaries that are to be sought. In this respect Britain is as well endowed as any part of Europe. Several of our estuaries are ranked among the most important along the European flyway. Morecambe Bay, that huge inter-tidal area between Cumbria and Lancashire holds almost a quarter of a million waders in winter. The Wash, the Ribble, the Dee and the Solway all hold over 100,000 birds.

Elsewhere Chichester Harbour has over 44,000 waders, many of which can be seen from West Wittering. Lindisfarne and its surrounding flats hold over 35,000 birds, while the Burry Inlet has almost as many. Most of these latter birds roost at the Whitford Burrows reserve, where they can be observed without disturbance. On the south coast the waders of Poole Harbour can be viewed at Brownsea Island, while the reserve at St Margaret's Bay, Kent, regularly produces a good collection of species. In north Norfolk the National Trust reserves at Brancaster, Scolt Head, Stiffkey and Blakeney Point are all good wader spots. In Arnold's Marsh at Cley, however, the Trust owns what is probably the best little wader marsh in the country – a marsh that is looked over by more bird-watchers every year than any other single spot in Britain. Access along the East Bank is free to all and it is a poor day when there are no interesting waders to be seen.

Two other little gems are worthy of note. In South Wales Oxwich Bay is an attractive haunt that should not be ignored by those intent on the huge concentrations of birds at the near-by Burry Inlet, while on the Isle of Wight the delightful Newtown Marsh is just the right size for comfortable viewing.

Waders may be largely ignored by the layman, but they offer one of the greatest of wildlife spectacles to be seen in Britain to anyone prepared to walk the lonely estuaries. In the places that hold them Britain has a special responsibility. So projects such as barrages, that threaten the very existence of these immensely rich habitats, should be examined in the greatest of detail before we rush ahead with so-called 'progress'. Our estuaries are held in trust for the nation.

Above left *Although waders, lapwings are mostly found on inland habitats, particularly farmland. They are a familiar sight following the plough, but many nests are destroyed by agricultural machinery.*

Above right *Four crouched lapwing chicks relying on camouflage to escape detection. Overhead the mother will divebomb any intruder and try to distract it from her young.*

Below left *Changes in agriculture, such as the introduction of modern machinery and insecticide sprays, have contributed to the near extinction of the stone curlew.*

Below right *During the breeding season the curlew moves away from the estuaries and mudflats to nest on moorlands.*

10 *Sea-birds*

ROBERT BURTON

The British Isles are generously endowed with sea-birds, and one of the earliest literary references to Britain's wildlife relates to sea-birds, in the Anglo-Saxon poem 'The Seafarer', written down in AD 1000 but probably dating back to AD 685. Being situated off the western seaboard of Europe ensures that sea-bird movement is intercepted, and the varied coastline of cliffs, rocks, beaches and estuaries makes these islands the most important place for breeding sea-birds in the north-eastern Atlantic. Twenty-four species of sea-birds breed regularly around the British Isles, and, although the largest colonies are on the coasts of Ireland and Scotland, examples of most of the twenty-four may also be found in England and Wales. Non-breeding species are seen as they travel along the coast on migration or gather to feed on concentrations of fish and other marine animals. Some fifteen species are regular passage migrants or winter visitors, with another twenty-eight as vagrants, rarely seen. As well as the true sea-birds, there are grebes, divers, and sea-ducks, which adopt a marine existence outside the breeding season, and waders, which haunt the shoreline to feed.

The best places for viewing sea-birds are usually headlands which jut into the broad path of the migrating birds, concentrating them into a narrow stream, as at Portland Bill, Dungeness, Blakeney Point and the Calf of Man.

FULMARS, PETRELS AND SHEARWATERS

The white body and grey wings of the fulmar can lead to it being confused with a gull at first sight, but it is a member of the tube-nosed group of birds to which the albatrosses, shearwaters and petrels belong. These birds are recognised by their tubular nostrils. They are oceanic, returning to land only for breeding, and in flight they have characteristic long, narrow wings, with which they alternately glide and flap, making use of air currents over the waves and around cliffs to gain extra lift.

A gannet carrying seaweed to build its nest.

Until a little over a century ago St Kilda, in the Outer Hebrides, had the only fulmar colony in the British Isles but the species has now spread to all coasts except the extreme south-east, and a few colonies have been founded inland. There are many National Trust coastal properties with fulmar colonies, ranging from St Margaret's Bay in Kent to Ramsey Island (over which the Trust holds protective covenants) off St David's. The fulmars arrive at their nesting sites in November and December, but they do not lay until May, after the pair has had a fortnight's honeymoon in which they go to sea and feed. No nest is made and the single, white egg is incubated for about fifty-three days. The chick spends a further forty-nine or so days on the nest, by which time it is heavier than its parents. It is fed on a rich diet of fish and other marine animals, which have already been partly digested to a sticky oil. Both adults and chicks defend themselves by spitting this oil a distance of about 1 metre (3ft) to deter predators. Colonies of fulmars are easy to spot because the birds sit on ledges and make loud cackling calls; the other British breeding tube-noses are by comparison retiring in the extreme. Manx shearwaters, storm petrels and Leach's storm petrels nest in burrows or under rocks, usually on remote islands, and enter and leave their nests only at night. The Manx shearwater is notorious for visiting the nest only on the darkest nights. All three species, together with passage species, are readily observed at sea but there are few breeding colonies in England and Wales. After a long absence, Manx shearwaters nest again on the Calf of Man in very small numbers and there are about a hundred pairs on the Trust's island of Lundy.

Despite the apparent confusion of a gannetry each pair of gannets has its own space, fiercely defended against intruders.

GANNETS

With the exception of Bempton cliffs on the Yorkshire coast, gannets nest on remote islands, and Bempton and Grassholm (off Wales) are the only British colonies outside Scotland. Nevertheless, gannets are wide-ranging and can be seen off any part of the coast, where their style of feeding makes spectacular viewing. The 2m (6ft) wingspan gives them the aerial grace of an albatross, but the easy gliding flight is transformed when a gannet spots a fish. It plunges vertically from a height of 30–40m (yds), with wings half folded, as it dives headlong into the water. The impetus of the drop carries the gannet underwater and within stabbing range of its victim, but it does not pursue prey. The head is protected from the force of the impact by a strong skull and a system of airsacs, while the nostrils open inside the beak to prevent water entering.

CORMORANTS AND SHAGS

Cormorants and shags are often seen perching with their wings outspread, apparently because their plumage gets waterlogged and has to be dried. If this is the case, a lack of waterproofing is an extraordinary shortcoming for such aquatic birds. They nest on rocky shores and islands.

These two species are easily confused at a distance, both being black, but the cormorant has a white patch under the chin and carries its beak tilted up when swimming. Both hunt fish by pursuing them underwater. They swim with their feet, keeping the wings folded. Most dives last for half a minute and the birds regularly descend to about 20m (yds). The cormorant's preference for flatfish in shallow seas and young salmon when it haunts fresh water has led to persecution by fishermen.

In Britain cormorants mostly nest around sea coasts but occasionally inland. They are fish-eaters and by the end of the breeding season their nesting colonies are fouled and evil-smelling.

Until recently gannets were an important part of the diet for many islanders round the Scottish coast. They were collected during the breeding season and salted for later use.

SKUAS

On the northern and western isles of Scotland and the furthest north of the mainland, there are colonies of great skuas, often known by the Shetland name of 'bonxies', and Arctic skuas. They attract attention by the vigorous way in which they defend their nests with dive-bombing attacks on intruders. Outside the breeding season skuas are seen around the coasts and they are joined by pomarine and long-tailed skuas from Arctic regions.

Skuas are relatives of the gulls and they are, similarly, basic fish-eaters, but the skuas have made their name as predators and thieves. The extent to which skuas attack other birds, either to kill them or force them to disgorge their food, is often overestimated, perhaps because it is such a spectacular sight. A sea-watch is greatly enlivened by seeing a skua in hot pursuit of an unfortunate gull, gannet, kittiwake or tern.

GULLS

These are the most abundant and familiar of all sea-birds. Nearly all are generalist and opportunist feeders and they lose no chance to exploit any source of food. Mainly coastal feeders that search the shores and shallow water for marine animals, they are less frequently seen out at sea except when following ships. To marine food, gulls add any offal or refuse, and several species have spread inland, where they feed on terrestrial animals – gulls following the plough have long been a familiar farmland sight. The larger species also turn predator, the great black-backed gull being a killer of rabbits and Manx shearwaters, and in some places expanding gull colonies are having an adverse effect on populations of smaller birds, even in nature reserves.

The black-headed gull is the smallest of our resident gulls and is immediately recognised by the chocolate-brown (not black!) hood on its head, which is replaced outside the breeding season by a dark spot behind the eye. Black-headed gulls are those most likely to be seen following the plough, and they have also spread into towns as scavengers. They nest in colonies whose sites range from saltmarshes and dunes, to hill lakes and gravel pits. The nests are close together and an intruder is greeted by a cloud of birds flying up and making harassing attacks, thereby drawing attention away from the three camouflaged eggs or chicks. Once the young have fledged, the colony is quickly deserted as it is a dangerous place for birds to linger unnecessarily.

The common gull is inappropriately named because, although it is widespread, it is not so abundant as other species and, moreover, it is easily overlooked. In appearance it is the size of a black-headed gull and has the general colouring of a herring gull but the grey of its back is rather darker, the legs are greenish-yellow instead of flesh-pink and it lacks the red spot on its bill. Its identity in a flock of gulls is often given away by its shriller calls, which have given rise to the alternative North American name of mew gull. Like black-headed gulls common gulls have increased in numbers this century and the two sometimes nest together in inland colonies. They also associate with tern colonies.

The herring gull is 'the seagull'. It draws attention to itself by its noisy, aggressive behaviour, quarrelling among its fellows and bullying other species. It is the principal seaside gull, where it scavenges both along the sea-front and in harbours of coastal towns and along beaches, where it searches rock pools and probes into the sand for anything edible. One trick is to fly up with shellfish and drop them from a height to smash their shells. This is not so intelligent as it seems, because a gull will drop its prey repeatedly and futilely on soft sand instead of seeking a hard surface.

An increase in numbers has led to herring gulls becoming a nuisance. Since the 1920s they have nested on buildings, causing noise and mess; they also foul reservoirs, and their habit of roosting on airfields makes them a serious threat to aircraft safety. They cause damaging erosion where they nest over burrows of puffins and shearwaters and they prey on tern colonies.

Despite a difference in appearance, the lesser black-backed gull is very closely related to the herring gull and hybrids occur naturally. The habits of the two species are very similar but many lesser black-backed gulls migrate to Portugal, Spain and north-west Africa for the winter. The greater black-backed gull is not only much larger, with darker wings, but it has pink legs compared with the yellow of the lesser black-backed gull. Although it feeds at sea more than most gulls, the greater black-backed gull is also more predatory during the summer. It catches rabbits and rodents and is a serious pest in Manx shearwater colonies.

One gull is very different from the rest: the kittiwake is a cliff-nester which leaves the shores completely after breeding and goes out to sea. The kittiwake has modified the basic breeding behaviour of gulls to overcome the dangers of nesting on cliffs. The nest, for example, is more solid in structure, with a high rim, than the nests of other gulls, and the chicks have sharp claws for hanging on. On the other hand, the nests are beyond the reach of terrestrial predators and camouflage has been relaxed. In recent years kittiwakes have also colonised warehouses and other buildings around docks and ports.

TERNS

Common and Arctic terns are virtually indistinguishable in flight, although the colour of the bill (the common tern's has a black tip) is a good aid to identification, as it is with little, Sandwich and roseate terns. The terns are delicately built relatives of the gulls, well deserving their old name of sea-swallows. They nest in colonies usually on the shore, but common terns nest on shingle banks in rivers and islands in sand and gravel pits. The main food is small fishes and crustaceans and a flock of terns feeding close to the shore is an entertaining sight. They dive headlong and either plunge or snap up their prey while barely touching the water.

A young herring gull. As they grow they wander increasingly far from the nest, but will be attacked viciously should they stray into the nesting area of another pair.

Even within the mêlée of a sandwich tern colony parent birds easily recognise their own chicks and feed them on small fish such as sand eels.

Despite a vigorous defence of the colony with dive-bombing attacks that culminate in an ear-splitting scream and often a thump on the head, tern colonies are very vulnerable because they often choose to nest in places favoured by holidaymakers. Even when housing and holiday building development have left the colony intact, the season's breeding can be destroyed by disturbance. Visitors should not approach too closely. Little and roseate terns now breed only in small numbers and protection for terns is needed, as at the National Trust's reserves at Scolt Head Island and Blakeney Point in Norfolk, where the public is excluded in the breeding season but hides provide a good view of the colonies.

AUKS

The auks have advanced further to a complete marine life than our other sea-birds. They come to land only to breed and they are proficient underwater swimmers. The wings are used for propulsion when chasing fish and the feet are used for manoeuvring. To this end the legs have moved to the rear of the body, so auks stand upright and waddle in an ungainly fashion like penguins. They have lost the ability to spring into the air so, to take off, they must launch themselves from a slope or run over the water to gain speed. The largest of the auks, the great auk, which was observed breeding on St Kilda in the seventeenth century, was extinct by 1844.

A visit to an auk colony often necessitates a boat trip, because the usual site is on a remote island or an inaccessible cliff. The four species choose different nesting sites. Guillemots gather in dense crowds on exposed cliff ledges or the tops of sea-stacks; razorbills prefer ledges and crevices. Black guillemots are hidden under boulders and deep cracks in cliff faces, and puffins dig burrows in the soil on the tops of islands or on cliff slopes. Lundy is perhaps the most famous breeding site in Britain for puffins, and the bird appears on the unofficial stamps.

When the auks return to their colonies in spring, the air becomes filled with circling black-and-white birds and the noise of their calls. All species lay a single egg except for the black guillemot, which has a clutch of two. No nest is made except by the puffin, which lines its burrow with grass. The guillemot's egg is sharply pointed so that it rolls in a tight circle and is less likely to fall over the edge.

The fledgling period varies widely between species. Despite the apparent safety of the cliff ledges, guillemots and razorbill chicks are preyed on by gulls and are accidentally knocked off by adults. These dangers are reduced by the chicks leaving the nesting ledge when they are eighteen days old and only half-grown. They descend to the sea on rapidly fluttering wings and swim out in the company of a parent. Black guillemots and puffin chicks are safe in their covered nests and fledge at about six weeks.

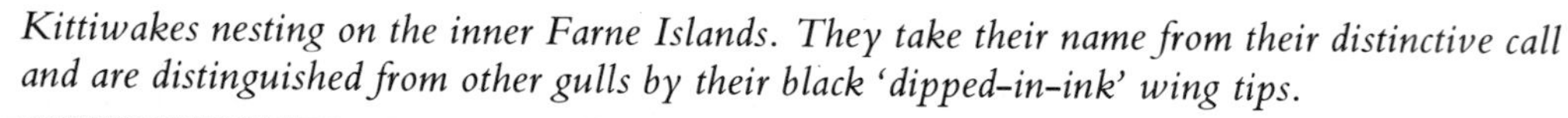

Kittiwakes nesting on the inner Farne Islands. They take their name from their distinctive call and are distinguished from other gulls by their black 'dipped-in-ink' wing tips.

After centuries of persecution for food or sport, British sea-birds have been able to make dramatic recoveries in numbers since the first Bird Protection Acts of the late nineteenth century. Apart from protection, increased food supplies in the form of offal from fishing boats and, in the case of gulls, adapting to new sources of food inland have played a part in these recoveries. Changes in climate may also have been important. Fulmar, gannet, shag and all the gulls have seen increases, sometimes dramatic, in their populations. Manx shearwater, most terns and auks have, on the other hand, shown declines, and the reasons are not always obvious.

There are several threats to sea-birds. The most striking, and the one which gets most publicity, is oil pollution. A major oil spill, like that caused by the wreck of the *Torrey Canyon*, results in huge numbers of oiled birds being recovered and no doubt large numbers being lost without trace. Auks are most at risk because they attempt to escape by diving through the oil, whereas other birds fly away. The effects of oil pollution on the size of population is not yet known, but a major disaster at the Scottish sea-bird colonies, where tankers pass within sight, must now be only a matter of time.

However, it is likely that other threats have an equal, or more serious, effect. It is not known, for instance, to what extent over-fishing is leaving sea-birds short of food. Pesticides, heavy metals and other pollutants may affect the breeding of birds even when they do not kill, and mass deaths have been caused from birds drowning in fishing nets. Modern nylon nets are not only difficult for the birds to detect but they do not rot and they continue to trap birds – and fish – long after they have been lost.

Such problems can only be resolved by government action, but the threat of disturbance at breeding grounds can be countered by the acquisition of reserves. Camouflaged tern eggs and chicks can be saved from carelessly trampling feet and the panics which send thousands of guillemot eggs and chicks raining into the sea can be avoided by keeping over-enthusiastic visitors away.

The National Trust's reserves are assisting in bringing peace and safety to sea-bird colonies, as at Strangford Lough in Northern Ireland and the Farne Islands, where there are colonies containing some of the few remaining roseate terns. The Trust manages the Farne Islands – first protected as a wildlife sanctuary by Cuthbert in AD 676 – so that 20,000 visitors each year can have the unrivalled pleasure of walking within a few feet of nesting terns, kittiwakes and cormorants. Incredible though it may seem today, had the Trust not launched the appeal in 1925 to save the Farne Islands, they would probably have gone to a shooting syndicate. The Trust also owns or protects a large number of coastal sea-bird colonies – in Cornwall alone, the Trust owns more than a quarter of the 300 miles of coast.

Left *Razorbills and their relatives, the guillemots, are the northern equivalents of penguins.*

Right *Shags, unlike their close relatives the cormorants, are confined to coasts and the sea and are rarely found on inland waters.*

11 *Birds of Prey*

IAN PRESTT

This group of birds includes some of the most magnificent in the world. Typically, they are large and powerful with keen eyesight and impressive powers of flight. With their talons they take mainly live prey, usually small mammals or birds, which are quickly despatched and dismembered with their short, sharply hooked bill. The food is swallowed whole or in large chunks, and undigested bones, fur and feathers are regurgitated later in smooth cylindrical pellets.

Birds of prey fall into two groups: the daytime hunting raptors – eagles, hawks and falcons – and the largely nocturnal owls. Of the raptors regularly recorded in Britain, four are widespread and common, four extremely rare and five are migrants. Five species of owl are relatively widespread and a sixth, the snowy owl of Arctic regions, is occasionally recorded, and has bred in the British Isles. Almost all the birds of prey can be seen on National Trust properties; some, such as barn owls, find Trust management of farmlands ideal.

RAPTORS

The hunting techniques of raptors are of three basic types. The large, broad-winged, short-tailed eagles and buzzards fly at considerable heights, interrupted by periods of circling or hanging almost stationary before they drop on their prey in a long dive or series of stoops. The slender-winged and long-tailed hawks and falcons engage in fast pursuit flights, twisting and turning dexterously to follow the prey's every movement. Harriers, by contrast, have a slow, buoyant flight and skim low over rough ground vegetation before pouncing on the surprised victim.

While the male broods their chicks, this female golden eagle stands guard on the edge of the eyrie.

The kestrel is the best-known British bird of prey, occurring in uplands, farmland, along coasts and even along motorway verges and in city centres. It hovers for long spells on rapidly beating wings and spread tail, hunting for the voles, mice and shrews which are often to be found in the grassy banks of motorways. In early spring displaying pairs can be seen circling high over their territories, calling with a shrill 'kee kee kee'. Kestrels do not build a nest, the four to six eggs being laid in a hollow tree, an old crows' nest, in a ruin or on a cliff ledge. They disperse outside the breeding season but most remain in Britain.

The three other British falcons do not hover, but capture their prey, usually birds, in flight. The largest is the peregrine, which is dark slate-blue above with pale, black-barred underparts; the males are only three-quarters the size of females. Its swift, powerful flight and devastating stoop single the peregrine out as one of the most formidable birds on earth. A hunting bird will hang, almost motionless, head to wind, high in the sky, before plunging at over 200kph (125 mph) on half-closed wings to strike or grab a bird flying beneath. Feral pigeons are favoured prey in Britain, but almost any bird, varying in size from a starling to a goose, may be taken.

Peregrines are highly territorial, and usually lay a clutch of four eggs on a cliff ledge or in the old nest of a raven or buzzard. This restricts breeding to coasts with sheer cliffs or mountain and fell regions, some of which is owned or managed by the National Trust. British peregrines are resident, but some move south in winter to feed on wintering flocks of waders and wildfowl.

The tiny merlin is hardly bigger than a mistle thrush; the brownish female is larger than the blue-grey male. During the breeding season isolated pairs are found in open upland habitats which usually include large areas of heather. The nest is usually a shallow scrape in deep heather, although old tree nests of crows may be used. Most merlins are resident within the British Isles, moving from the uplands to winter on coastal plains and estuaries, where they are also joined by merlins from Iceland and other northern breeding

Although outlawed, pole traps are still responsible for the deaths of many owls and birds of prey.

A female kestrel. This is the most common British falcon, nesting even in the centre of cities, and is found in almost every habitat in the countryside.

areas. Here, they feed on flocks of finches, pipits, larks and small waders.

The remaining falcon – the hobby – is a migrant, returning to Britain from Africa during May. The British breeding population, which probably does not exceed a hundred pairs, is largely confined to central southern England, although pairs are occasionally seen elsewhere. A high-speed flier with long, slim wings, it hawks for large dragonflies, and other large insects and can overtake martins and swifts in full flight. It usually nests in old tree nests of crows.

The only other raptor as widespread as the kestrel is the sparrowhawk. However, living in woodlands makes it much less evident. As with many other raptors, the slate-grey males are smaller than the brown females. Sparrowhawks feed mostly on woodland birds, which they surprise with a short dash from a perch or take unawares as they round a bush or hedge. Sparrowhawks are usually seen when they circle overhead, often mobbed by small birds, or as they move from one wood to the next. Conifers are their favourite nesting trees, where a large, larch-twig nest is built against the trunk amongst the lower green foliage.

The goshawk, which is like a larger version of the sparrowhawk, is still rare. It was exterminated earlier this century and it was not until re-introduction (often by birds which had escaped from falconers) and recolonisation in the late 1960s that once again it started breeding regularly.

The common buzzard is both widespread and locally abundant in upland regions of western and northern Britain and in the valleys of the south-west. It is a large brown bird with a majestic soaring flight, during which it will sail effortlessly in large circles on fully spread, rounded wings and tail. Often several will circle together, exchanging their plaintive mewing call. In spring circling pairs proclaim their territories and build stick nests in trees or on cliffs, where they usually lay a clutch of three eggs. In Scotland and the Lake District they can be confused with the larger golden eagle.

Another buzzard, a migrant occasionally found wintering in parts of southern Britain, is the rough-legged buzzard, which breeds in tundra around the Arctic circle. Less closely related is the honey buzzard, a rare summer visitor of which fewer than twenty-five pairs nest in a few scattered southerly locations where sufficient nests of the wasps and bees on which they feed can still be found.

Though its numbers are increasing, the red kite is rare – fewer than fifty carefully protected pairs remain. It occurs in central Wales in a habitat which it shares with the common buzzard. Although the kite's habits are similar to those of the buzzard, its diet includes more mutton carrion, earthworms and insects. It is easily distinguished from the buzzard by its larger, thinner wings and extremely long, deeply forked tail.

All three harriers are uncommon. The most numerous and widespread is the hen harrier, which now numbers about 750 pairs. It is a moorland nesting bird, nesting on the ground and laying its eggs in a shallow scrape. Several of the brown plumaged females will mate with one male and in spring the superb sky-diving display of the silver-grey males is a fine sight on the open moor. The marsh harrier is less common (about twenty-five pairs in Britain), and as the name implies it is a bird of wetlands, building its nest on the ground in reedbeds. Montagu's harrier is a summer migrant closely resembling the hen harrier in appearance and habits, which is now almost extinct in Britain as a breeding bird; it is found mainly on heaths and lowland moors, although in common with the other two species outside the breeding season it is also found hunting over marshes, estuaries, water meadows and even farmland.

Until the end of the nineteenth century there were two species of eagle in Britain, the white-tailed eagle and golden eagle. The former was exterminated by man and the latter greatly reduced. The golden eagle is a large powerful bird, able to survive harsh highland winters and to breed at altitudes of up to 500m (1,625ft). It feeds mainly on mountain hares, red grouse and ptarmigan and rears a single, occasionally two, young in large cliff or, less commonly, tree nests.

Under protection, the golden eagle numbers are slowly increasing and it has recently even nested in England. The Nature Conservancy Council is currently attempting to re-introduce the white-tailed eagle on to the Rhum National Nature Reserve.

Above *Peregrines were among the most popular birds for falconry; this one has taken a magpie. Unfortunately unscrupulous collectors still rob nests despite strict protection.*

Below *A female sparrowhawk sheltering her well-grown young from the rain.*

The osprey is Britain's only fish-eating bird of prey.

The osprey returned to Britain as a breeding bird after a gap of half a century. Over twenty pairs now breed each year in Scotland with pairs rearing two or three young in large, conspicuous tree-top nests. The osprey feeds almost exclusively on fish and is a dark brown bird with a conspicuous white head; it hovers high over the water before diving and plunging with outstretched talons to grasp a fish just below the surface, which it then carries off head-first to eat at a favourite perch.

OWLS

The British owls are medium-sized, plump-looking birds, with large heads, flat faces and forward-looking eyes. Their specialised sight enables them to see in almost total darkness and their acute hearing helps them to hunt at night. Their plumage is soft and dense, producing almost silent flight and this enables them to surprise their prey, which they hunt by flying low and buoyantly over the ground. Small mammals form the principal prey of most owls, but insects and roosting birds are also taken. Like the day-flying birds of prey, they regurgitate pellets compounded of the undigested fur and bones from their prey.

The bulky, conspicuous nest of the osprey made it easy for gamekeepers and egg-collectors to exterminate the species by the turn of the century. However, under strict protection, their numbers are now increasing year by year

The tawny owl is the most widespread and common owl and is found in forests, woods, copses and even in suburban gardens. It is strictly nocturnal, but, like the other species, you may see it at dusk, perched on a fence post or sitting on a house roof near the chimney where it is warm. It passes the day roosting, often standing against the trunk in an evergreen tree or shrub. It is usually heard rather than seen, when it utters a wavering hoot or sharp 'kewick'. When hunting, tawny owls wait on a low branch and drop on to their victims, which are usually swallowed whole. Tawny owls nest in a hollow tree or the old nest of a crow; they also readily nest in artificial nest boxes.

Long-eared owls are similar in size to the tawny, but appear greyer and slimmer and have prominent feather tufts resembling ears. Their eyes are bright orange. They usually roost and nest in conifer woods or thick hawthorn scrub but hunt in the open, sweeping back and forth on long wings on the darkest of nights. They breed in old nests of sparrowhawks, magpies and other woodland birds, laying four or five eggs in March. The numbers of long-eared owls have decreased, and although very widely spread they are now quite scarce.

The short-eared owl closely resembles the long-eared, but it has very different habits, often hunting by day on open fells and moors, where it nests on the ground. The field

Left *The diminutive little owl is less strictly nocturnal than other owls and is often seen on telegraph wires, gateposts and other prominent perches in broad daylight. It usually nests in tree holes, but occasionally in rabbit burrows, as here.*

Right *By day tawny owls roost hidden away in thick cover. Should small birds discover one they will mob it, presumably to draw attention to the presence of a potential predator.*

vole forms its main prey and the numbers of short-eared owls in an area will increase during the years when vole populations reach 'plague' proportions. In March displaying birds, diving and clapping their wings high over open ground, provide an exciting spectacle.

The fourth relatively common owl, the barn owl, contrasts with the three previous species both in appearance and habits. The upper parts are pale golden buff and the breast white, giving a ghostly white appearance as it suddenly appears out of the darkness in car headlights. It is a bird of parkland and farmland, roosting and nesting on ledges or straw bales in barns, derelict buildings or hollow trees. During hard weather, or when the adults are feeding young, barn owls often hunt by day.

The two remaining owls are the little owl, no bigger than a blackbird and the largest, the snowy owl. The former was introduced to Britain from the Netherlands at the end of the nineteenth century and is now thinly distributed throughout England. The latter bred successfully in the Shetlands in the late 1960s and 1970s and is an occasional visitor.

Opposite *The ghostly white barn owl, which often nests in buildings such as church towers, and which calls with an alarming shriek, undoubtedly gave rise to many tales of hauntings.*

HABITAT AND CONSERVATION

The numbers of birds of prey and their distribution within Britain have been markedly affected by a long history of harsh persecution and continuing habitat loss, especially in the lowlands. It is for these reasons that many, particularly the larger species, are relegated to remote, inhospitable regions. Thus golden eagles are mainly confined to the mountains and glens of the Highlands; the common buzzard, which at one time bred throughout the British Isles, is now primarily a bird of northern and western uplands; and the remnant of the population of our once common red kite is confined to central Wales.

Their destruction was largely dictated by game interests and to a lesser extent by fishing and farming concerns. The development of efficient and cheaper sporting guns and greater accessibility to the country led to a tremendous expansion of sporting estates. To manage these the owners employed large numbers of gamekeepers and water bailiffs who, with relentless determination, destroyed everything with a hooked bill and claws. Guns, poisons and an incredible array of traps were used. As the predators' populations declined, their rarity value increased and the skin and egg collectors intensified their efforts to obtain the last specimens. The white-tailed eagle, goshawk and osprey were all wiped out and many other species reduced to isolated remnants.

It was not until well into the twentieth century that effective bird protection laws were finally introduced in response to growing public concern over the failure to safeguard our diminishing wildlife and because of an increasing interest in the countryside. This is clearly demonstrated by the tremendous growth in membership of such bodies as the National Trust and the Royal Society for the Protection of Birds since the Second World War. There can be no doubt that this surge of public opinion and more effective legislation have been major factors in recoveries now taking place in our bird of prey populations. Red kites have increased from a remnant of a few pairs to over thirty regularly breeding pairs. Two pairs of golden eagles are now firmly established in the Lake District and hen harriers have spread from survivors in Orkney to breeding localities as far south as Wales.

While an increasing number of estate owners and farmers take an interest in conservation and show greater toleration towards birds of prey, a regrettably high number of incidents is still uncovered every year. Too many estates continue to use poletraps although they were made illegal at the turn of the century. Protected birds of prey are all too frequently shot and their nests destroyed. Another continuing and probably increasing problem and one that is proving extremely difficult to bring under control is the illegal use of poison in the countryside. Although put out to control foxes and crows, it is indiscriminate and many protected birds of prey are killed. Another recent problem, now largely brought under control in Britain, arose from the widespread introduction into post-war agriculture of certain persistent organochlorine insecticides such as DDT and dieldrin. These compounds, now controlled through a voluntary Pesticide Safety Precaution Scheme, proved particularly harmful to birds of prey at the end of the food chain, killing large numbers and rendering many others sterile by causing eggshell thinning.

It is regrettable that at the same time as birds of prey are beginning to benefit from a more sympathetic attitude, the loss of habitat is causing growing anxiety. Birds of prey need large hunting territories to provide food throughout the year. Some of the smaller species are able to survive in the lowlands following the transformation of these areas from woodland to farmland, but further mechanisation and increasing intensification of agriculture is telling against them. The transformation of herb-rich permanent pasture into cereals robs the barn owl of its traditional hunting sites, and the eradication of hedges, banks and small spinneys to create large fields robs the kestrel of its prey. The surviving pockets of semi-natural common land, such as the southern lowland heaths so favoured by the hobby, are steadily being built over or altered.

Even in remote northern and western uplands, which previously provided refuge for many species, the situation is changing. Conditions in these inhospitable regions were never easy, and many birds of prey found it hard to survive during harsh spells in winter or to find the extra food needed to feed young during the early summer. But even here modern techniques promote change and reduce the remaining feeding and nesting sites. Heather moorland, necessary to hen harriers, merlins and short-eared owls, is steadily being transformed to grassland or buried under conifer plantations. Alongside these changes the construction of better roads and the increase in the number of family cars means that greater numbers of visitors from towns and cities are opening up the surviving areas of wilderness.

The time is past when the future of our birds of prey can be left to chance. Continuing vigilance over law enforcement and increased tolerance will be needed more than ever to help those species that can exist in the environs of man, and more nature reserves and other protected areas must be established in the countryside to help more sensitive species.

OPPORTUNITIES FOR OBSERVATION

The extensive variety of woodland, rough fields, moorland and stretches of coast owned and managed by the National Trust provides valuable refuges for our birds of prey and present excellent opportunities for viewing them, thereby adding to the enjoyment of a day in the open air. The meadows and parkland surrounding many of the Trust's country houses and ruins provide excellent hunting grounds for the kestrel and barn owl. The mature garden, shrubberies and copses adjoining others will harbour tawny owls.

The Trust's impressive series of cliff sites along the south coast of England, round the south-west peninsula and along the Welsh coast enables one to see coastal nesting peregrines among the colonies of sea-birds. Moorland properties, such as those on Dartmoor, Exmoor and in north Yorkshire and Northumberland, hold breeding merlins, hen harriers and short-eared owls. A visit to a coastal marsh in East Anglia or to an inland fen such as the Trust's long-established holding at Wicken in Cambridgeshire may provide a glimpse of a hunting marsh harrier or an opportunity to see a little owl hunting at dusk from a fence post or pollard willow. Walkers on Trust land in Cumbria may be rewarded with fine sights of buzzards, and in the valleys with the occasional glimpse of a hunting sparrow hawk, or possibly with the unexpected find of a long-eared owl roosting in a larch or Scots pine or a rare sighting of a golden eagle. Indeed, few of the Trust's holdings in the countryside will be without one or more species of birds of prey.

In their nest in the middle of a reedbed these young marsh harriers are beginning to test their wings. Although once on the verge of extinction in Britain, their numbers have increased slightly in recent years.

PART THREE

Other Wildlife

12 *Amphibians and Reptiles*

KEITH CORBETT

Amphibians and reptiles represent two major groups of vertebrates, 'primitive' perhaps in evolutionary terms but well adapted and successful none the less. Both types are essentially cold-blooded and depend therefore upon outside temperature to attain the warmth necessary to support their daily active life. It is not surprising that Britain with its cool, temperate climate has relatively few species.

NEWTS

We have in Britain three of these tailed amphibians, the palmate, the smooth or common and the great crested or warty newt. All live on land but return to water each spring to court, mate and lay up to several hundred eggs, usually curled individually into aquatic plant leaves. The males of each species display a breeding livery of crests and tail flashes, the latter being an integral part of the complex tail-waving courtship.

The males leave the water in late spring while the females remain and continue egg-laying until at least midsummer. Unlike our frogs and toads, the newts will feed under water as well as on land, typically on any small invertebrates and dead animal matter that they come across. Newts are rarely seen on land save for the few found resting under stones and logs, usually in the vicinity of the breeding pond. The great crested appears to require dense scrub near to the pond or at least woodland cover within 50m (yds) or so, but this does not seem to apply to the other species.

During the summer months, the four-legged and tufty-gilled larvae (the newt tadpole or 'eft') grow and progressively absorb their gills to emerge eventually from the water with their land-stage coat of 'velvety' skin. This is the species' dispersal stage and during the two to three years before maturity they may find another local pool. Just how far they wander in this time is not yet known, but the smooth newt's frequent colonisation of new

Previous page *Six-spot burnet moths feeding on thistles.*

During the breeding season the male common newt develops a wavy crest and bright colours on the underside, and performs elaborate displays to attract females.

ponds in gardens certainly suggests that it is more successful than the other two species.

On bright sunny days in the spring, one may see newts rising briefly to the surface to gulp in air before diving out of sight, but the best way to observe (and identify) them is after dusk when they are most active. Walk around the edges of ponds with a powerful torch and look particularly at the shallow margins where the newts are very much more visible and will usually carry on courting despite the light.

Identifying the newts is easiest in springtime when the populations are concentrated and the male's courtship dress is evident. The smooth and great crested both have a dorsal crest but that of the great crested is higher and more spectacular, and it alone has the warty, dark brown to black skin on back and sides. In all ways it is the largest of our newts and not just in its length of 14–19cm (6–8ins), but also in sheer bulk. The palmate, often speckled beautifully with greens, is the smallest species at around 5cm (2½ins), and its male boasts only a low ridge as a crest, but it has dark webbed feet as the identifying breeding feature, from which it gets its name, and a spikelet end to its tail.

Although all three species can be found together in some pools, the great crested and smooth avoid the acid conditions of moor and heath which so often support abundant palmates.

It seems doubtful that the palmate really prefers such poor waters even though it thrives in them; it is more likely that it can only flourish in the absence of competition

Left *Frogs have disappeared from large areas of rural Britain with the filling in of most farm ponds. However, suburbia with its numerous garden ponds provides an ideal habitat where frogs can flourish.*

Right *The natterjack toad is easily distinguished from the more widespread common toad by a bright yellow stripe down its back.*

and predation from the other two species which actually require richer water conditions. The palmate may attain dominance again as ponds in richer soil areas begin to sour and acidify from too much tree shade and leaf litter. Deeper water is sometimes quoted as another requirement of the great crested but perversely it happens that the two large populations in south-east England are in water respectively 5m (18ft) and 43cm (18ins) deep.

Conservation problems for these species derive from the widespread loss, pollution and neglect of ponds; also the 'park syndrome', wherein the short-cropped grass gives little cover or food for either adult or emergent young, and where too many waterfowl (encouraged by feeding and pinioning) can add to pollution, remove all aquatic weed and may even feed directly on newts and their young. Equally damaging is the introduction of predatory fish, though even too many goldfish can be a cause of serious competition. The tiny but pugnacious stickleback can be a less obvious but major problem, with few newts able to survive their harassing unless there is good water-weed cover.

FROGS AND TOADS

There are three species of native tailless amphibians in the British Isles, all very different in habits and requirements from each other. The common or grass frog is angular in shape, smooth, damp-skinned and rarely strays from moist habitats, often spending much of its terrestrial life amongst the deep grasses, herbs and mosses of the pond margins. Its skin overlies rich spongy tissue which enables it to 'breathe' when underwater by the direct diffusion of oxygen, and this in turn allows it to hibernate at the bottom of ponds. When they are available, stream- or spring-fed pools are particularly chosen for their supply of oxygenated water; without this in a stagnant pond they are in danger of drowning if the water freezes over for a long time in the winter as the oxygen is used up by micro-organisms and is not replaced through the ice barrier.

Depending upon the timing of spring, spawning can be as early as January, when frost and snow are still around. Hibernating in or close to their spawning pond, the whole frog population's arrival is fairly synchronised and spawning can be completed within a few days. The males croak gently to attract the female and then clasp them from behind in the 'nuptial embrace', their grip aided by darkened, horny nuptial pads on the thumb and forefingers. The end product is the spawn clump, often laid *en masse*, and which by the rapid absorption of water forms the familiar tapioca-like blobs known to every one who has ever visited a pond in spring. Many thousands of eggs are produced per clump, which is just as well because the tadpoles are potential prey for any number of vertebrate and invertebrate predators until the day they leave their pond. As they grow, the tadpoles go through the classic amphibian metamorphosis; from external to internal gills and then to lungs, developing first the back and then the front legs, with the tail re-absorbed as they climb on to land as a young froglet. Those that then survive terrestrial predators may return as adults to breed within two years. On land, their only defence is their long back legs and their leaping ability, and the occasionally used 'scream'. This latter ploy is reserved for 'mortal peril' situations and apparently so startles a would-be predator or

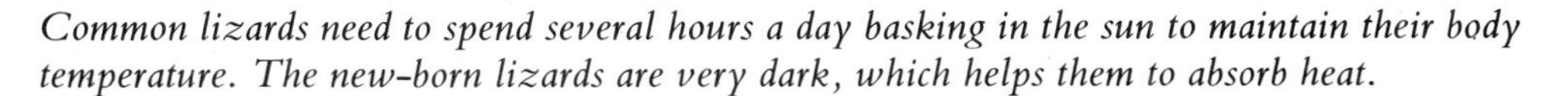
Common lizards need to spend several hours a day basking in the sun to maintain their body temperature. The new-born lizards are very dark, which helps them to absorb heat.

Common frogs in amplexus. *The males, usually smaller than the females, arrive first at the ponds, where they lie in wait, and may embrace the females even before they reach the water.*

trampler as to allow just enough time to leap to freedom. However, it has been most frequently heard when the frog is about to be swallowed by a grass snake. Unfortunately for the poor frog it can then have absolutely no effect, because in common with all snakes, this predator is completely deaf! Herons are notorious predators on frogs in the spring and in common with other birds they will eat all except the female's oviduct. These are rejected and dropped below the perching or nesting tree or in flight, and their contents soon absorb moisture to swell into jelly lumps and become the 'star-slime' of country folklore.

Our common and natterjack toads are both rounded in shape with dry warty skin. They also share a unique defence: the presence of complex poisons which are readily secreted from certain of the skin glands or 'warts' on their head and back when faced with, or seized by, predators. This makes them both distasteful and poisonous to warm-blooded birds and mammals, which usually leave these species alone. Some however, notably the crow and magpie, can learn to adapt their feeding to the non-toxic undersides and simply disembowel them to eat the innards.

Common toads return to ancestral breeding sites, which can involve them in migrations

Left *Common toads return to the same pond year after year to breed.*

Right *A common lizard.*

of over a mile, including crossing roads. Should the lake or pond be close to a road or pathway then the males' behaviour may increase the risks of accidental injury or death, as they will use this as a focus to waylay incoming females and ensure that they are already in nuptial embrace before even getting to the water. Clasping and mating is similar to that of the frog but the period can be stretched over several weeks and their spawn is in the form of a continuous 'spaghetti-like' string. These are wound around vegetation or branches in mass twinings and in deep water, but the natterjack tends to lay singly and on to the bare sandy floor of warm, shallow water.

The penetrating nocturnal call of the male natterjack toad attracts others to the nearest suitable pool. It is a sound which, as a chorus, can be heard over a mile away and is not dissimilar to that of the nightjar. Spawning is later than that of the common toad, starting in mid-April, and is sometimes protracted into June. The natterjack has a characteristic yellow to cream stripe down the centre of its back and rather short sturdy back legs used for digging burrows as daytime and winter retreats. These powerful legs enable it to run rather than to hop as the common toad does. Surprisingly for an amphibian, the natterjack likes to bask in the sunshine, although both toad species hunt and feed largely as nocturnal animals. The natterjack is restricted in Britain to sand dunes and dry heathland, however it is almost extinct now in the latter. The National Trust land at Frensham was one of its last breeding sites in Surrey, but it became extinct in the 1970s, possibly due to human disturbance since this site was a particularly popular recreation area for Londoners.

In addition to the native amphibians, the Alpine newt, marsh and edible frogs, midwife toad and some others have been introduced. The only species which flourishes is the marsh frog, which is abundant on Romney Marsh, Kent, particularly in the Royal Military Canal – 5·5km ($3\frac{1}{2}$ miles) of which (near Appledore) are owned by the National Trust.

LIZARDS

In Britain and north-east Europe the sand lizard is restricted to sandy habitats because of its need for an open and warm soil in which to lay and incubate its 5 to 15 whitish eggs with parchment-texture shells. They are laid around mid-June, and hatch from mid-August to late September, hunting and feeding until their hibernation at the end of October. Young hatchling sand lizards have the patterns of their adult's distinctive ocellate spots, which will in later life blend superbly with the flowers and seed-heads of the deep heather in which they live; the few surviving populations isolated in the north-west English dune system also have these basic spots, but their edges are usually coalesced into a more stripy pattern in harmony with their marram grass habitat. The young sand lizards reappear with the females in mid-April. A month or so earlier the males will have emerged to take on their brilliant green (from emerald to lime) breeding colour after their first slough (moult). At this time local rivalry develops, though this normally results in many more chases and submissions than actual combat.

Both the sand lizard and the much smaller and more lightly built viviparous or common lizard incubate their eggs within the female's body by basking, but the sand lizard lays her eggs in soil, whereas the common retains hers up to the point where laying becomes almost synchronised with hatching. Freed from the need to find specific egg-laying sites, this species is able to exploit many open habitats such as rough grasslands, hedgerows, bramble and herb banks, as well as the dry heath and sand dunes necessary for the sand lizard and is also able to ascend the moorlands and mountain slopes. The heaths of Surrey, Hampshire and Dorset were, until the middle of the nineteenth century, extensive. Even within the last quarter of a century the few remaining heaths have been fragmented or destroyed through afforestation, agriculture, suburban encroachment and increasing numbers of caravan sites. The National Trust owns a number of important heathlands, several of which have breeding populations of sand lizards and other reptiles.

Britain's only other native lizard, the slow worm, is legless (and is neither slow nor a worm!). It feeds mainly at dusk, dawn and during or after rain, for its main food consists of slugs and worms, with small white slugs being favoured. Although slow worms bask, they usually hide away under bark or stones. The slender 7cm ($2\frac{3}{4}$ins) long young are a striking gold colour with a dark stripe running the length of the body. The adult females retain this pattern but the gold blurs to a browny bronze; the males lose the stripe to become more uniform, and with age some individuals may have azure blue spotting.

All British lizards can shed part of their tail at will. This enables them not only to escape when seized by the tail, but the wriggling and writhing broken end may also distract and confuse a predator by presenting a moving alternative prey; the number of surviving lizards with shed tails bear witness to the success of this ploy. The tails regrow – but are never quite as perfect as the original.

Above *The adder is Britain's only poisonous snake, but its bite is unlikely to prove fatal. Senseless persecution has exterminated it in many areas, but it is still abundant.*

Below *The smooth snake is confined to a few isolated colonies in southern England. Despite strict protection it is still threatened.*

SNAKES

Unlike lizards, snakes have no external ear openings and no closable eyelids – hence the 'unblinking and impenetrable stare of the snake'. Like lizards, they 'smell' by flicking out their forked tongue and returning scent molecules to a taste organ in the roof of the mouth. As they are relatively high up the food chain, all British adult snakes feed on other, often quite large, vertebrates. They are able to do this by 'dislocating' the jaw and increasing the gape for swallowing.

The adder or viper is characterised by the prominent bold zig-zag pattern all down its back, and has venom glands which are able to pump lethal poison into its prey via two hollow hypodermic-like teeth (the fangs) – lethal that is to the relatively small prey animals such as voles and mice but hardly ever lethal to man. To put this into perspective, many hundreds of people have succumbed to bee stings in this century, but only half a dozen or so to the adder. Most bites are provoked by attempts to catch the snake; there really is no justification for the needless persecution inflicted upon all snakes (and even the harmless slow worm) on the grounds that 'it might be an adder'.

Whilst the adder is short (usually less than 55cm [2ft]) and stocky, the grass or ringed snake, with its typical light-coloured collar, can grow up to 110cm (4ft) and occasionally 145cm (5ft) in length. The smooth snake is more slender and rarely exceeds 48cm (20ins) and lacks any collar or zig-zag. The grass snake feeds mainly on amphibians and fish but the smooth snake is a specialist feeder on other reptiles, particularly the lizards, although it will also eat nestling birds and mammals should it find them. It is virtually confined to lowland dry heaths, which are one of our richest reptile habitats and the only one in which six species can be found. The grass snake was once abundant in the low wet meadows of Britain. Most of these lush, flower-rich hay meadows were in the floodplains of rivers, but they have mostly been drained and ploughed.

After springtime mating, the pregnant females bask and incubate their developing young. The adder and smooth snake give birth in August and September–October respectively. The egg-laying grass snake has had to adapt to our cool climate by seeking out rather artificial habitats in which to incubate its eggs. Typical choices include compost piles, dung heaps, and sawdust waste. Unfortunately the first tends to be insecure, and the latter two are becoming obsolete. The young have a strong instinct to return as adults to these ancestral and often communal nesting sites, and this can sometimes involve a migration of over 3km ($1\frac{3}{4}$ miles). The threats to and the gradual disappearance of these incubation sites could soon result in a collapse of this species in many parts of Britain, and it would seem sensible to try to locate and protect some of the surviving sites if not to re-create others. Before man provided these 'incubators', grass snakes were probably dependent on the regular natural flooding of lowland river valleys to deposit piles of rotting vegetation. This too has become a thing of the past as we have progressively drained and controlled our lowland rivers.

CONSERVATION

To ensure these species' survival and the naturalist's and public's continued enjoyment of them, sympathetic land management measures have now to be considered. It is here that land-owners and managers like the National Trust have to discharge carefully their very real responsibilities for our natural heritage in balancing these animals' special needs against the sometimes potentially conflicting pressures on their properties.

With the disappearance of many old ponds and the draining of marshes the obvious answer to many of the problems is the restoration of old ponds and the creation of new. This type of management has already been undertaken on National Trust properties, particularly where survey information was available for guidance as on the South Downs, north Kent, and the Dorset coast. Higher priorities may be assigned when declining species like the great crested newt and the natterjack are involved. Here Enterprise Neptune, the Trust's coastal acquisition programme, has shown its wildlife worth; pool excavation in the Merseyside dunes has been very successful, while on the Norfolk coast a sandy pond initially designed for terns has proved to be a bigger attraction for the natterjacks, and more ponds may be planned. The provision of a new breeding pool and the necessary re-introduction of this species is under consideration for the Trust's heath properties in Surrey.

This latter area also contains important sites for local sand lizards and smooth snakes and a very detailed management programme is well on the way to completion. Management for reptiles is interlinked with the essential need to keep habitats open and relatively free of scrub. In the special case of heathland, itself an endangered habitat type, it has also to be protected against accidental fire and the encroachment of ever-present pine and of fire-induced birch scrub and bracken.

Sadly, one of each of the discussed groups of species has declined to such rarity and is under such threat that it is now strictly protected under the 1981 Wildlife and Countryside Act: the great crested newt, the natterjack toad, the sand lizard and the smooth snake. The 1980s could prove decisive, not only in provision for these rarer species but also in ensuring the natural abundance of the so-called commoner species, most of which are now in decline and one of which, the grass snake, is giving rise to serious concern.

13 Fish

ALWYNE WHEELER

Compared with many animals of the countryside fish are not often seen and are generally little studied. Most visitors to National Trust properties which contain water habitats will, unless they are anglers, at most see the flash of a silvery flank in the water, although they may be lucky enough to witness a trout leaping to a surface-living insect, or in a few places a salmon lurking in a pool. Carp can sometimes be seen cruising beneath the surface, and on occasion will become very tame and join waterfowl in coming to feed on bread thrown into lakes by visitors. Shoals of minnows can be glimpsed darting beneath bridges, but such glimpses are, however, only the visible fraction of a great wealth of life under-water. A glance at the contents of an angler's keep net will often reveal what is hidden in a lake or river. Worldwide there are at least as many species of fish as there are mammals, birds, reptiles and amphibians added together.

Because of their sheer wealth of numbers fish are very important in freshwater or marine habitats, apart from their value to man as food or for recreation by angling. Their place in the scheme of life is principally as predators on invertebrates and other, smaller fish (for virtually all our native fish are predatory), and as prey for other aquatic animals and birds such as the kingfisher, heron, merganser, and otter in fresh water, and in the sea terns, cormorants, puffins, seals, porpoises and dolphins.

FRESHWATER FISH

There are fewer than forty species of freshwater fish found in Britain, but they inhabit virtually all rivers and lakes in the country, the only exceptions being a few small lakes and streams high up mountains and those rivers and lakes suffering from extreme pollution; fortunately these are relatively few compared with a century ago. Mountain streams offer a rather special habitat; the water is fast-flowing, cold and well oxygenated. These factors also limit the invertebrates that can live in such streams. As a result there is frequently not much food available for the fish, and those species that inhabit such streams are small, or are small specimens of larger species. In such hill streams as are found in the Lake District, the Yorkshire hills, Snowdonia, and the Highlands of Scotland the dominant fish are trout, stone loach, bullhead and, rarely, minnow, although some do not occur in all these areas. Both the stone loach and bullhead live concealed under stones and in crevices and thus escape the force of the current. Brown trout in such streams are small, perhaps growing to a maximum of 20cm ($8\frac{1}{2}$ins) at an age of several years, and because of their size are able to hide in crevices, while the biggest specimens are able to monopolise such pools – often only the size of a wash basin – as offer quiet water and slightly better living space. Trout from such habitats are frequently beautifully coloured, dark bodied

Salmon fishermen setting nets near Cotehele Quay on the Tamar.

with olive bellies and a brilliant scattering of red and yellow spots. The minute trout from streams like those on Dartmoor and Exmoor, as well as the northern hills, are often so bright that it is difficult to believe that they are the same species as the big silvery trout of neighbouring lakes.

Some of Britain's most fascinating, and beautiful, fish live in the lakes of Snowdonia, the Lake District, Scotland and Ireland. These are the charr and the whitefish. All are the descendants of fish which penetrated the lakes soon after the last Ice Age, migrating to spawn in freshwater from the sea in the case of the charr, and from the great ice lakes that bordered the Arctic ice in the case of the whitefish. With later changes in sea level relative to the land, and the disappearance of the ice lakes, populations became isolated in the British lakes and their river systems. Today, the whitefish are represented by two, perhaps three, species in the British Isles; one is found in Llyn Tegid in North Wales, where it is known as the gwyniad, in Haweswater and Ullswater, where it is called schelly, and in Loch Lomond and Eck, where it is the powan. This whitefish has a rather pointed snout and the lower jaw is shorter than the upper. The vendaces are found in the Lake District in Derwentwater and Bassenthwaite, as well as in Castle and Mill Lochs at Loch Maben in Dumfriesshire in south-east Scotland. Vendaces have the lower jaw protruding beyond the upper. Both species are olive green on the back with strikingly silvery sides and belly –

Brown trout vary considerably both in size and colouring. The smallest and most brightly coloured are those living in fast-flowing upland streams.

hence whitefish – and have the small, fleshy, adipose fin on the back near the tail fin which shows they are related to the trout. The closely related pollans are found in loughs in Shannon and other parts of Ireland.

By contrast, the charr is much more widely distributed, being found in about a hundred lakes in northern and western areas. It is also, at spawning time at least, vividly coloured, the males having a brilliant pinkish-red belly, with white edges to the ventral fins. The Snowdonia charr, found in Llyn Peris and Padarn, is known as torgoch, literally 'red belly'. In past times charr were important food fish locally, the Windermere population being particularly well known in the form of potted charr. They still are fished for in small numbers both in Windermere and in a few other lakes.

The interest in both the charr and the whitefish populations is that, having been isolated in their lakes for perhaps 15,000 years, each population has responded to local conditions by evolving from the ancestral forms. As a result fish from different lakes often look rather different from one another.

Although the lakes in the highland areas of the British Isles are noted particularly for their charr, whitefish, and trout stocks they also contain other fishes, such as pike, perch, and in the shallow water minnow, bullhead, and stone loach. The real wealth of fish life, however, is seen away from the hills in the lowland rivers and lakes of central, eastern and southern England. Because of its proximity to the continent eastern England was the receiving area for all the purely freshwater fishes (as it was for many other animal and plant groups). Finding their way through the flooded lands of what was to become the North Sea, where the river Thames once joined with the Rhine, freshwater fishes colonised England after the last Ice Age through the river systems between Yorkshire and Kent. As a result there has always been the greatest number of fish species in this area, although the species distribute themselves in the rivers according to their requirements for water quality and flow.

Thus, upstream in the river Dove, where it flows through Dovedale in the Peak District, the trout and grayling find the cool, fast-flowing water greatly to their liking. The grayling is a spectacularly beautiful fish with a large dorsal fin, which when the fish are spawning in spring has a pale orange edge and dusky spots with a violet tinge. It grows to about 50cm (21ins) in length, is a distant relative of the trout, and when fresh is said to smell faintly of thyme. At spawning time it is not at all shy and can be seen in clear water swimming in pairs in water of 1m (3ft) depth, but it occurs only where the water is of good quality.

Further downstream in a typical river the flow becomes slower, but is nevertheless still moderate; the water is often cloudy, slightly warmer and may contain less oxygen. In such an area the dace is abundant, as is the roach in slow-flowing pools; occasionally pike lurk in the weed beds of the pools, and minnows form enormous schools. The freshwater eel occupies crevices in the river bed hidden from the current, and makes forays mostly at night in search of invertebrates on the bottom or any dead fish near by. It is in this part of the river that the salmon will breed after migrating from the sea, cutting their spawning redds (nests) in the finer gravel of the river bed and burying the eggs to preserve them from predators of all kinds. Here the eggs lie through the worst of the winter to hatch in April or May, becoming active as fry a month later when the minute crustaceans, on which they will feed, are at their numerical peak.

In the lower reaches of rivers other fish are dominant in the slow-flowing, muddy-looking water. Where riffles form the dace is abundant, but in the pools its relative the chub is more common. Below weirs in the broken water barbel occur, but they are also common on clean gravel bottoms in the main river. In deeper water the roach and the bream form large schools, the latter feeding close to the bottom on buried insect larvae and worms. Eels are often abundant, especially in autumn when the maturing adults are migrating towards the sea and eventually their breeding areas near Bermuda; while in spring millions of nearly transparent elvers, young eels, will be moving up-stream, already about 10cm (4ins) long, three years of age and veterans of a transatlantic crossing.

Not surprisingly, the fish that thrive best in still waters are those species which live in these lowland rivers. Lowland gravel pits, reservoirs, and lakes are usually full of roach and bream, with variable numbers of carp and tench, and perch and pike. Still waters

Migrating salmon were once a common sight in nearly all the rivers of Britain, but as a result of overfishing and pollution they have disappeared from most.

such as the pits at Wicken Fen in Cambridgeshire are the ideal habitat of the rudd and in the overgrown edges of the pits the ten-spined stickleback can be found. The Fens are also the major habitat of the small silver bream, which can be caught by the thousand in the Lode at Wicken. This area was the habitat of the burbot, a fish which is now extinct in Britain. The only freshwater member of the cod family, it grew to over 1m (3ft) long and a weight of 5kg (11lb). It was at one time found in a number of rivers between Yorkshire and Cambridgeshire.

No one really knows why the burbot became extinct in Britain (it is still common in most of northern Europe) but it was probably a combination of factors such as local pollution, possibly overfishing in places, and changes in the habitat. Too many fens were drained and slow-flowing rivers dredged clear to improve the flow and drainage, with the result that habitats suitable for the burbot became fewer and further spaced. Other fish which have restricted ranges are also vulnerable to changes in their habitat. Some populations of charr have been exterminated, like those in Ullswater by pollution from the Greenside lead mines in the late 1800s. The Castle Loch population of vendace is also extinct, killed it is thought by changes in the loch resulting from the discharge of sewage effluent from the town of Loch Mabern and by competition and predation from introduced native fish.

The redistribution of native fish is one of the greatest threats to fish of restricted distribution. The charr, trout, salmon, and whitefish of the highland areas have now to survive in the company of introduced predatory pike and perch in almost all their known habitats, and in many lakes they have to compete for food with other introduced species like roach, dace, chub, minnow and bream. Even more damaging is the introduction of exotic species, particularly predatory fish such as the zander, which was released into a river in the Great Ouse system and which has now spread, and been spread, through much of eastern and southern England. There is now little hope of controlling it and it will probably continue to increase in range. As it is a fish-eating predator, particularly well adapted to survive in lowland rivers, one of the consequences of its introduction is a great reduction in the numbers of the smaller fish, for example roach, on which it feeds.

Much of the redistribution of native fish has been made in the alleged interests of angling, but the results have not always been favourable to angling. The stocking of waters is also a common fisheries management practice, made in earlier years with little regard to the compatibility of fish and habitat, or to the possibility of spreading parasites and disease. Fortunately with recent legislation such transfers of fish can only legally be made with the permission of, and after tests have been made by, the appropriate regional Water Authority, so this undesirable habit can now be controlled.

There are, however, exceptions to the general rule that all exotic species are harmful. The carp, introduced from central Europe in medieval times, has now become a valued part of the fauna of many lowland lakes and rivers, and lends itself to semi-domestication in moats, fish ponds and ornamental waters. The rainbow trout has also proved to be a 'model' introduction. Now reared in hundreds of fish farms around the country it is stocked in rivers, lakes and reservoirs, but it rarely breeds naturally and despite its abundance there are only about a dozen self-sustained populations known in Britain. Again, the rainbow trout is almost domesticated in certain artificial waters and will come to the surface to be fed.

SEA FISH

The carp and the rainbow trout are two of the visible fish that the bankside observer may see, but there are few such species in the sea. One fish that is from time to time visible near the surface is the basking shark. Even though it is the largest European fish (and the second largest in the world), reaching a length of 11m (36ft), it often comes fairly close inshore and can be seen from the cliff tops, especially in Cornwall and the Isles of Scilly, swimming slowly just below the surface, its large, triangular and floppy dorsal fin and tail tip jutting out of the water. It is quite harmless as its food is the minute animal plankton of the surface, and it can be approached closely in a rowing boat, but it is an immensely powerful fish and has to be treated with caution.

Generally the only other time it is possible to see marine fish in their natural environment is on rocky shores where there are pools. Here, by lying flat on the rocks patiently, a number of species of shore fish can be lulled from their alarm and will emerge from hiding to behave normally. Absolute stillness is essential though. Rocky shores, such as that at Wembury in south Devon, contain a wealth of fish which is rarely guessed at on first impression. The most common species there, and on most other rocky shores in Britain, is the shanny, a slender, dark green or brown fish with its eyes placed high on the sides of the head and a pair of finger-like pelvic fins under its belly on which it will prop itself while searching for food. It is usually up to around 10cm (4ins) long. This is about double the length of Montagu's blenny, which occurs on the south and western coasts, from Dorset to Pembrokeshire, and on the Irish coast. Montagu's blenny is a bold little fish which will emerge from hiding quite quickly. It lives in small pools on the shore about midway between the extremities of high tide and low tide, usually in shallow basins which have purplish coral-like growth of red seaweeds on the rocks. It is very distinctive as it has a fringed triangular flap of skin above the eyes, and small pearly-blue spots on the sides of the head and body. Breeding males have the edge of the triangular flap and the angle of the mouth bright orange; it lays its eggs in a crevice of the rock in July.

A common shore fish at Wembury, and elsewhere on rocky shores around the coast, is the little worm pipefish, a 10cm (4ins) long dark brown, stick-like fish. Like other pipefishes the male carries the eggs on his belly; one can often find 'pregnant' males in late summer. Although common it is extremely hard to see, as by colouring and body form it is a perfect mimic of brown seaweed and to a lesser extent the egg-wrack. So competent a mimic is it that it never moves quickly – to do so would give the game away – but it lies close to the stem of the weed, its tail loosely curled around it for anchorage. This is but an extreme example of a rule amongst shore fish that they all use coloration to match the rocks or weeds amongst which they live in an attempt to outwit predators. This is seen in the long-spined sea scorpion, which may be deep red or dull green depending on whether the individual fish is living in red or green seaweed.

One of the exceptions to this rule is the giant goby, which makes only a mild attempt to match its colouring to its habitat, but then it is larger than most shore fish being about 25cm (10ins) long. It will lie quite openly on bare rock in pools near the high-tide mark, but is exceptionally wary and any sudden move by a pool-watcher will send it darting for cover. It is a shore fish which is strongly associated with National Trust coastal sites, for the furthest east it extends up the Channel coast is Wembury, while it occurs at Polperro, Mevagissey, Porth Scatho, and virtually all the way along the south Cornish coast, and in the Isles of Scilly.

Gobies are the most successful group of shore fish and one species or another can be found in almost all marine coastal habitats. Mostly small fish, they have broad heads, chubby cheeks, and eyes placed well up on the side of the head. Underneath, the pelvic fins are united together into a disc, by means of which they can prop themselves up clear of the sea bed and increase their visual range. In sandy habitats, like those on the north Norfolk coast, near Scolt Head and Blakeney Point, wherever there is standing water near the low-tide mark the pools will contain sand gobies, small sandy coloured fish which dart away when disturbed only to merge again with the background on settling. In summer they breed prolifically and their young become important food for tern chicks, as well as the adult birds, which also eat large numbers of sand eels, another abundant inhabitant of sandy shores. These shores are also the nursery grounds for the dab and plaice, whose postage-stamp-sized young live in the breakers, but close to the sea bed.

In estuaries another community of shore fish can be found. Here the most abundant goby is the common goby, which lives both in the mouth of rivers and in saltings pools. Rarely longer than 6cm (2¼ins) in length, it abounds in summer and autumn, following earlier successive spawnings, the eggs being laid on the underside of cockle and other mollusc shells. In places like the Essex Blackwater estuary and the Cuckmere estuary in Sussex this goby is probably the most abundant fish, but the low shore is also the habitat favoured by flounders, and bass and grey mullet are found commonly in deeper water. Just as the keep nets of riverside and lakeside anglers are a good source of information about which fish are present, so on the coast the catches of anglers fishing from the shore or a pier provide an indication of what fish are in the locality. The catches of coastal fishing boats are also always worth a look and species rarely seen in the fishmonger's – such as garfish and mullet – can often be examined.

So specialised are the fish found on our coasts that any disturbance to the local environment leads inevitably to their decline. The building of marinas in estuaries and along the marshy shores converts a special habitat to a wildlife wasteland. Development of holiday resorts near rocky shores quickly leads to the degradation of the habitat for the fish, other animals, and plants that depend on it for survival. The shoreline, be it rocky, sandy, or mud and marsh, is a very vulnerable habitat which it requires constant attention to safeguard. The National Trust has taken under its care much coastline of the south-west of England and thus offers protection for some of the most interesting habitats for shore fish; other areas, such as the north Norfolk coast, are also protected in this way but there is so much ecologically valuable coastline that these areas are small by comparison with those that are still threatened.

A full-grown minnow with two young tench.

14 *Invertebrates*

Dr M. G. MORRIS

There are about twenty times as many invertebrate species of animals in Britain as the rest of the animals described in this book put together. Despite the vast array of species and great abundance of individuals they have to be searched for, since most invertebrates hide themselves away and only a few voluntarily make themselves seen, or felt. Terrestrial invertebrates include mainly worms, for instance the familiar earthworms of the soil; the jointed-limb invertebrates, or arthropods, which include insects, spiders and crustaceans; and a third group, the molluscs, which include the slugs and snails.

Although most slugs and snails are very small and inconspicuous, some are spectacular; they are most likely to be seen after rain, on a warm summer evening. The Roman snail, increasingly rare, is found in dense vegetation in chalky areas. Its shell grows to up to 5cm (1¾ins) across, at which size the animal may be nine or more years old. Although Roman snails have a wide distribution in Europe, they are described in the *Red Data Book* (which lists endangered and threatened species) as rare. For centuries they have been exploited for food, and not only in France – there are traditional recipes for snails in cider from Gloucestershire and the Cotswolds. However, exploitation for food, particularly in France, has led to their depletion, and a large international trade has developed which could threaten them further. As long ago as 1909 declines were being noticed in England. For anyone wishing to eat snails, the common garden snail is equally edible. The remains of black and yellow striped snails, together with other common species are often found in heaps around a stone or on a wall – these are the so-called thrush's anvil where a song thrush has smashed the shells open.

The largest slug, the great grey, grows to over 15cm (6ins); like snails slugs leave silvery trails which trace the routes of their nocturnal wanderings. Besides the numerous and varied insects, the arthropods include such groups as woodlice, centipedes and millipedes, and the spiders and their allies. The woodlice include some very abundant species in grassland and woodland as well as some species with restricted ranges; several are particularly associated with humans and human artefacts, such as gardens and buildings.

Centipedes feed on other small animals, whereas most millipedes feed on dead plant material. The millipedes include the cylindrical forms and the flattened and sculptured types as well as usual species such as the pill millipede, easily mistaken for a woodlouse. Woodlice, centipedes and many other invertebrates are most easily seen by turning over stones or logs – but always remember to replace these gently in exactly the same spot.

There are about 600 different species of spiders in Britain, many of them very small and rare. Their habits are varied, although all spiders are predators. In addition to the familiar orb-web spinners, there are species which make small, hammock-type webs which are spread horizontally on the ground; some species do not make webs, including the wolf spiders, jumping spiders and many others. Crab spiders lie in wait for prey inside flowers, which they often closely resemble in colour, while the attractive ladybird spider – red, black and white in the seldom-seen male – is one of the rarest species, protected by the Wildlife and Countryside Act. Another rare species is the aquatic raft spider, which takes its name from the erroneous belief that it could use leaves as a raft. The commonest spider is the so-called garden spider, which makes a beautifully symmetrical orb-web. It is easily recognisable by the white cross-like marking in the middle of its back.

The harvestmen (daddy-long-legs) are superficially similar to spiders, but they lack the division of the body into separate head-thorax and abdomen; confusingly the name daddy-long-legs is also sometimes used for craneflies, which are true flies. The harvestmen are often particularly abundant in old meadows. Also related to the spiders are the few British species of pseudoscorpions, small (mostly under 0·5cm [¼in]) animals with scorpion-like pincers but lacking the scorpion's tail and sting.

Left *A banded snail.*

Right *Black slugs emerge at night to feed. They in turn are preyed on by hedgehogs.*

Harvestmen or daddy-long-legs are neither spiders nor crane flies, both of which they resemble superficially, but a separate group of arachnids.

The most frequently seen invertebrates are the insects, which can be placed in three main groups: the primitive wingless species; those which develop directly from an active immature or nymphed stage into the winged adult; and those which have a generally inactive pupal (chrysalis) stage between development from a larva (caterpillar, maggot or grub), which is very unlike the adult, to the mature insect.

The spring-tails of any damp grassland, woodland or wetland are examples of primitively wingless insects. So are the bristle-tails of coastal cliffs and heathlands. The group also includes a few other more or less obscure types.

Most familiar of the insects without a pupal stage are the grasshoppers and their allies (orthopteroids), or the aphids, depending on whether one is a country-lover or gardener! Many grasshoppers are common, but a few are rare and threatened. Most of the short-horned species are grassland or heathland species. The meadow grasshopper is found particularly in lush, tall grasslands, whereas the common field grasshopper prefers dry, shorter swards. The long-horned grasshoppers or bush crickets are mostly found in shrubs and include our largest orthopteroid, the great green bush-cricket. The oak bush-cricket, which is often attracted to light, is the only tree-living British species and is also unusual in being almost entirely predacious instead of feeding on plants. The orthopteroid insects include other groups such as the groundhopper, crickets and mole crickets; several of them are rare and threatened. Even when they are not visible, they are often heard, each species producing its own distinctive chirping 'song'.

Related to the notorious aphids are the plant bugs, the leafhoppers, planthoppers and froghoppers, the jumping plant lice, the white flies and scale insects, altogether a very varied series of insects. The plant bugs include plant-feeding froghoppers, which produce 'cuckoo spit', the predacious terrestrial forms such as shield-bugs and, less commonly than in previous ages, bed-bugs, but also many familiar species of ponds and rivers, such as pond-skaters and water-boatmen which can be seen on almost any pond or slow-moving stream. The insects with a complete development include dragonflies together with four large groups of insects and some smaller, and perhaps less familiar, ones. The four groups are the butterflies and moths, beetles, bees, wasps and ants and two-wing flies.

Below left *dark brown bush cricket;* below right *crab spider eating a fly;* bottom left *cardinal beetle;* bottom right *wasp beetle.*

Butterflies are among the most conspicuous, as well as the most popular insects, and are considered separately (Chapter 15). The division into butterflies and moths does not have any real scientific basis; for instance the hawk-moths differ from certain other moths as much as they do from butterflies. In Britain the easiest way of distinguishing butterflies from moths is by the antennae: butterflies have knobbed or club-ended antennae, moths are variable but not knobbed. There are many more different species of moths – some 200 – usually divided (quite unscientifically) into the large ones and small ones. Virtually all feed on plants and many are very local, with some, such as the Essex emerald or New Forest burnet, very rare and endangered. The largest species, the hawk-moths, are spectacular.

The caterpillars are often as striking as the adult butterflies and moths, and in some cases their behaviour is of particular interest. The caterpillars of the large blue butterfly relied on an association with ants, which took the caterpillars into their nest to milk honey-dew from them; the caterpillars fed on ant grubs. Unfortunately, largely owing to the disappearance of the short-turf grasslands with wild thyme (on which the caterpillars also fed) the large blue recently became extinct in Britain. Many species of butterfly and moth caterpillar have very particular requirements for food plants, and are often only found on one or two species. For instance, most of the coppers, including the large copper, feed on dock.

Some of the moth caterpillars are extremely striking in appearance. In the case of the cinnabar (which unusually for a moth often flies by day) the caterpillars have a bold black and yellow striping and are often very visible feeding on ragworts. Their vivid coloration is a warning to would-be predators of their extreme distastefulness – to the point of being poisonous to most animals. Many caterpillars have a 'furry' skin, which is often very irritating to potential predators – but some predators have overcome this and the cuckoo, for instance, is almost a specialist feeder on woolly-bear caterpillars.

Just as many butterflies and moths, such as the eyed hawk-moth or peacock butterfly, have 'eyes' on their wings which can suddenly be flashed so as to frighten predators, some caterpillars have spectacular displays. The pussmoth enlarges its red and black head and waves its long 'tails', and the elephant hawk-moth caterpillar has 'eye' spots which can be distended. But one of the most common forms of defence is camouflage. Some caterpillars mimic twigs, and so do some adult moths, such as the buff tip, which at rest resembles a broken twig. One of the most unexpected defence mechanisms is that of the death's head hawk-moth. It is a surprisingly large moth, and if seized by a predator it emits a loud squeak – certainly astonishing enough to make most humans let go immediately!

Mimicry among insects is rife – not all flying bee- or wasp-like insects are bees or wasps. Many harmless insects have evolved to mimic bees and wasps because of their stings so that would-be predators leave them alone. There are numerous flies which look just like wasps and there are also moths and beetles which mimic bees.

There are nearly 4,000 British species of beetles and they have very varied habits.

Most of the ground beetles are predacious, as are many rove beetles. There are several different aquatic families including carnivorous great water beetles which will eat frogs and small fish, and tiny whirling beetles and plant-feeders such as the crawling water beetles. Many of the rarest species are associated with old forest remnants, such as Sherwood, Windsor and the New Forest. There is a wide variety of species of terrestrial plant-feeders, such as the leaf beetles and weevils. Beetles also include some notorious pests; the most famous of which, the death-watch beetle, is a major problem in many National Trust properties.

The familiar bumble bees, wasps and ants are allied to an enormous array of species which are virtually all parasites of other insects. The most spectacular of the wasps is also the most frequently persecuted: the hornet. The hornet has declined markedly in recent years, but despite its reputation it is peaceful if not molested. Also included are the fascinating gall wasps, one of the many groups of insects and other animals which induce plants to form galls, usually in a form characteristic to each animal species. The gall wasps attack oak predominantly, and include such familiar galls as the oak apple and oak-marble gall. The domestic bee, along with many other insects, plays an important part in pollinating crops and wild plants. Domestic bees, together with other species of bees, wasps and ants often live in complex social groupings. The richest habitats for bees and wasps are found in the south of England, and are often associated with heathlands of Surrey, Dorset and Hampshire. Wicken Fen is also of considerable interest.

In the *Red Data Book* all five species of European wood ants are listed as vulnerable. Some countries, such as Switzerland, have given legal protection to them, but as yet there is no formal protection in Britain. Where they occur, usually within or along the edges of forests, wood ants' nests are usually very conspicuous, with a large bare mound of pine needles and twigs surrounded by an area devoid of vegetation. Some species of ant keep aphids which they 'milk' for sweet secretions.

The two-winged flies are probably the most varied order of insects, with terrestrial

A wood ants' nest in a chestnut plantation.

Cinnabar moth.

and aquatic species, as well as predatory, plant-feeding and parasitic, and a range of forms with bizarre shapes, habits and life histories. Many, such as mosquitoes and houseflies are pests to humans or, like onion fly and carrot fly, their crops. Hover-flies, midges, horse-flies, bee flies, fruit-flies are a few of the 5,200 odd found in Britain.

About forty-two species of dragonflies and damselflies are found in Britain. Several of these species are now rare, or endangered, with a few others already extinct. Their disappearance is due to the enormous loss of suitable wetland habitats. Not only has drainage and pollution taken its toll but the run-off of fertilisers and removal of farm ponds, particularly in the last quarter of a century, has caused a rapid destruction of Britain's dragonfly fauna.

Inevitably, then, much of our varied invertebrate fauna is conserved when any land is protected. The National Trusts, unwittingly in some cases, conserve a large part of this interesting and relatively little-known fauna. Nevertheless, special attention must be given to rare and endangered species, particularly those of threatened habitats, such as wetlands and woodlands, or those endangered by changing agricultural and forestry practices, such as the replacement of semi-natural grasslands with ryegrass leys, or the abandonment of coppicing. Fortunately, insects are often resilient creatures and their small size and high fecundity often mean that their conservation can be achieved in relatively small reserves.

In the recently published *Red Data Book of Invertebrates*, of the eight species listed in Britain five are freshwater animals. Of the two species of freshwater mussel, one (Spengler's freshwater mussel) is already extinct, and the freshwater pearl mussel is listed as vulnerable. It was exploited for its pearls – many of which appear in regalia and European

Above *Lime hawk moth.*

Below *Mating damselflies.*

Above *Tiger moth.*

Below *A darter dragonfly.*

crown jewels, and in Britain it is still threatened by over-collecting; it is also very sensitive to pollution. Mussels may live for up to a hundred years, and do not mature until they are about twelve to fifteen years old. Consequently, once there is any depletion of a population it takes an extremely long time for them to recover. Two species of freshwater crayfish are listed in the *Red Data Book*; the noble crayfish is considered vulnerable and may exist in Britain as a result of introductions. Its disappearance from Europe is largely as a result of a crayfish disease which has swept Europe since the 1860s. The white-clawed crayfish is listed as rare, but one of its last European strongholds appears to be Britain. However, should the commercially important American crayfish ever be introduced, they could soon wipe out the native species, since as yet the crayfish plague has not entered Britain. Finally the medicinal leech is listed in the *Red Data Book* as being of indeterminate status – no one really knows if it is endangered with extinction, but it certainly needs conservation measures. Until this century medicinal leeches were used in hospitals in their hundreds of thousands and even now they are used in small numbers in biomedical research. Medicinal leeches live under stones in ponds and streams, particularly in areas where there are grazing animals. The leech may not be everyone's idea of an endangered species, but its importance as a producer of anti-coagulant hirundin demonstrates the importance of conserving all wildlife – even invertebrates – however insignificant or seemingly repulsive.

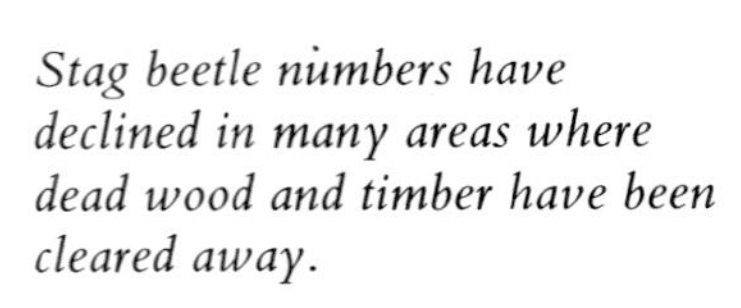

Stag beetle numbers have declined in many areas where dead wood and timber have been cleared away.

15 *Butterflies*

Dr M. G. MORRIS

Although Britain has only some sixty species of butterfly, these few give pleasure to many hundreds of entomologists, general naturalists and ordinary visitors to the National Trust's open spaces and nature reserves. Although some species are restricted to particular habitats, many are widely distributed and are even found in gardens and parks. Besides the notorious large and small whites, whose caterpillars devour cabbages, many species are attracted to garden flowers, particularly buddleia and Michaelmas daisies. Therefore, in the grounds of many of the Trust's country houses with extensive flower gardens, 'aristocrats' such as small tortoiseshells, red admirals, peacocks and occasional commas and painted ladies can be seen in abundance, especially in late summer. They are often joined by the yellow males and pale-greenish females of the brimstone. Another 'garden' species is the holly blue, whose caterpillars feed on the buds of holly, ivy and other garden shrubs.

However, most butterflies are to be found in the countryside away from towns and villages. We have eleven species of 'browns' including such familiar species as the ubiquitous meadow brown and small heath, the wall, hedge brown or gatekeeper and ill-named marbled white. Our eight species of fritillary are all tawny-orange in colour and many of them have silver spots on the undersides of the wings. The red admiral and the white admiral are both related to the seldom-seen purple emperor. The little Duke of Burgundy fritillary is in fact related to the blues, coppers and hairstreaks, with eight, two and five British species respectively. The 'cabbage whites' and brimstone are listed with the wood white, orange tip in the 'white' family together with three immigrant species of clouded yellows. Our splendid swallowtail (restricted to East Anglia) is our only representative of a family with nearly 600 species world-wide. Finally, we have eight species of rather small, dull, but fast-flying butterflies, the skippers, which are less closely related to all the other butterflies.

Despite the small number of butterflies which inhabit Britain, most habitats support one or more species. The little mountain ringlet is truly montane, flying in the Buttermere mountains of the Lake District and the Scottish Highlands south of the Great Glen, but

Left *The pearl-bordered fritillary. Fritillaries are fast-flying, beautifully marked butterflies. The markings are often similar to those of the flower, the snake's head fritillary.*

Right *Grizzled skipper.*

seldom at elevations of less than 2,000 feet. Agriculturally unimproved grasslands (several of which are owned by the National Trust) support many of our species. The chalk downs, such as Box Hill, support the adonis and chalkhill blues and the silver-spotted skipper, besides many of the more common species of browns, and dark green fritillary, the common and small blues, which are not so restricted to calcareous grasslands. Most of the skippers are also grassland butterflies; the dingy and grizzled skippers appear in late spring to be followed by the large skipper in midsummer and the small and Essex skippers in late summer.

Although butterflies are sun-loving insects, many occur in woods and some can be regarded as butterflies of scrub. The hedge brown and ringlet are examples, though the ringlet is also common in woods. The green hairstreak is another species which frequents shrubs and thickets. Most of such butterflies are found in deciduous woods only, but the wood white has adapted to the rides of conifer plantations and is often common provided the rides are neither too wide and sunny nor too narrow and closed-in. Fritillaries, such as the silver-washed and pearl-bordered, are often common in appropriately managed oak woods in southern England. Other woodland butterflies include the shade-loving speckled wood, the purple hairstreak (whose caterpillars feed on oak and are, perhaps, more easily seen than the adult butterflies), and the glorious purple emperor, more common than it seems but not often seen. The white admiral is much more frequently seen as it glides expertly in and out of the branches.

Although extinct in Wicken Fen, the swallowtail survives in several areas of the Norfolk Broads.

The pretty little silver-studded blue and the grayling are our only real heathland butterflies. The grayling also occurs in coastal habitats. The marsh fritillary is one of our few wetland species, although not entirely restricted to marshes. The swallowtail, now confined to the Norfolk Broads, and the re-established large copper are also wetland butterflies.

The caterpillars of butterflies are much more difficult to find than the butterflies themselves and even more difficult to identify. They are mostly well camouflaged and secretive, hiding on the underside of the leaves of their food plants. The choice of food plants is very wide, and although many species are found on a wide variety of plants, they often have preferences. The coppers are often found on dock and sorrels, the browns on grasses, the blues often on trefoils and related plants, many fritillaries like plantains and violets, and the red admiral, comma and peacock all like stinging nettles.

Despite popular belief, most butterflies live at least a week or two, and several species, such as the peacock and small tortoiseshell, hibernate as adults and therefore may live as long as nine months. Some butterflies have two broods each year, with the second brood spending the winter as pupae which emerge in the spring.

In addition to Britain's resident butterflies, a number occur as migrants, some coming from surprisingly far away. On a summer Channel-ferry crossing it is not unusual to encounter butterflies (and other insects such as ladybirds) mid-Channel, but the twenty-one miles from France to England are only a fraction of the total journey of some species. The most spectacular is the monarch butterfly, which at very irregular intervals manages to get to Britain – from America. The Camberwell beauty from southern Europe is of slightly more regular occurrence, and has also occasionally appeared in large numbers. As might be expected from its name it is a magnificent butterfly. It probably comes as quite a surprise to most people to learn that the red admiral and the painted lady, although quite common British butterflies, are probably unable to survive British winters. Some hibernating adults survive, but the bulk of the British populations seen on the wing in August onwards derive from eggs laid by butterflies that migrated from southern Europe earlier in the summer. Other species of British butterfly which are migrants or have their populations occasionally boosted by migrants include the clouded yellow, the small white, green-veined white, the large white and the small tortoiseshell.

The intensification of agriculture and forestry has destroyed many colonies of butterflies. Draining, ploughing and re-seeding, together with clear-felling and the planting of coniferous monocultures have been very damaging, but more insidious changes, such as the use of herbicides and fertilisers and abandonment of management practices

Peacock caterpillars spin themselves a fine web to give some protection from predators.

The meadow brown is one of Britain's most widespread and abundant butterflies. These two are mating and their wings are rather frayed – violent rainstorms or hail can wreak havoc with a butterfly's frail wings.

such as coppicing, have also caused damage. Ryegrass leys and mature sitka spruce plantations are not suitable for any of our butterflies. Habitat destruction and alteration are the most important factors in the decline of our butterfly fauna. By comparison pesticides and over-collecting, although serious, are relatively insignificant, but acid rain and, in the long term, climatic changes (which cannot be reversed) may also prove to be important.

A few butterflies have become extinct in Britain within the last two centuries. In addition to the large blue, which became extinct by about 1980, the mazarine blue had disappeared from most of England by about the 1880s. The black-veined white, although once fairly widespread, had disappeared by about 1930.

The swallowtail population declined and disappeared from much of its former habitat in the fens when they were drained. At one time there were some 2,500 square miles of fen stretching from Lincoln to Suffolk. Now almost all that remains of that great fen is the Trust's property at Wicken, the first part of which was acquired in 1899. The swallowtail butterfly is now extinct in Wicken Fen but survives in the Norfolk Broads.

The large blue, now extinct in Britain. However, attempts are being made to reintroduce it.

A pair of mating small tortoiseshells.

A peacock.

A comma (left) *and a red admiral.*

Another fenland speciality was the large copper butterfly, which probably became extinct some time in the 1840s, when the last specimens were caught in Holme Fen. The large copper is widespread and often abundant on the Continent, and in 1909 some caterpillars were imported and liberated at Wicken Fen. However, this attempt was a failure because the caterpillar's food plant – the great water dock – was uncommon. In 1915 a population of large coppers was discovered in Friesland, in The Netherlands, which was very similar to the extinct British race; it too was rare, and in 1927 a small number were introduced into Wood Walton Fen Nature Reserve, where under intensive management it has thrived ever since. During 1929–30 great water docks were planted at Wicken Fen and large coppers were subsequently liberated and successfully colonised the Trust's reserve, though many were lost when this part of the fens was ploughed during the Second World War.

Other butterflies are under threat. Few colonies of the heath fritillary remain. Those of the adonis blue have declined by half every twelve years since 1950. On the other hand, appropriate management has virtually guaranteed the survival of the black hairstreak.

Monitoring and management research can provide the answers to a range of conservation problems, but action needs to be taken. The single most important activity for conservationists if the protection and appropriate management of land. In this the National Trust is playing a leading role. The main site where it is planned to re-establish the extinct large blue, now that the area has again been made suitable through management, is owned by the National Trust. Nearly two-thirds of the larvae colonies of the very local Glanville fritillary are on Trust property.

Occasionally the Trust earns a bonus: one of the reasons for acquiring Fontmell Down was the large colony of chalkhill blues there. Only after acquisition was it discovered that the site supports the largest colony in Britain for the much more local silver-spotted skipper, a species reduced to about fifty colonies in Britain.

16 *Seashore Animals*

JOYCE POPE

Forty-five per cent of the finest coastlands of England, Wales and Northern Ireland is already owned by the National Trust and, totalling some 692km (430 miles), it includes a wider variety of habitats than any other sort of holding. Almost every type of seashore is rich with life, although this is often not immediately apparent, for the majority of animals on the beach are small and many of them belong to groups which do not occur on land. Some may be familiar as animals of the open sea, but the beach fauna is not the same as that of the oceans and truly marine creatures are doomed if they are cast ashore.

Every part of the coast, from high, wave-battered cliffs to broad, sandy, dune-backed beaches or mud flats, has to contend with a greater daily range of conditions than those of any other sort of environment. The reason for this is the tides, which twice every twenty-four hours are dragged by the forces of the moon and the sun towards the land, then pulled oceanwards again. At their greatest amplitude round the British coasts they may reach a height of more than 10m (33ft) but it is usually less than this. However great or little it may be, the water movement leaves the upper part of the beach uncovered for much of the time, while lower down the beach plants and animals are inundated for most of their lives. So the shore is zoned, with organisms living in the upper part able to stand desiccation while the animals found lower down depend on being almost permanently water-covered.

The lives of all seashore creatures are affected by more than mere wetness or dryness for many other conditions of life change with the presence or absence of water. On a hot, windy day rocks and sandy surfaces heat up to dangerously high temperatures and the water in small rock pools may be uncomfortably hot even to the human hand. Yet this warmth disappears with the splash of the first wave of the incoming tide, which may bring water 20°C cooler, so animals which live in such places must be able to stand great and sudden changes of temperature. As the water in a rock pool heats it is less able to hold dissolved oxygen, and as it evaporates it becomes more saline. During the day, when plants are photosynthesising, the water tends to be alkaline, but at night when they produce carbon dioxide as a waste product it is markedly acid. These variations in temperature, salinity and pH are far greater than those which could be borne by land-living creatures, but above all the shore dwellers must be able to withstand the power of the waves, which in times of storm come crashing in to smash all but the best protected. Some, therefore, tend to carry heavy shells which can withstand battering and rolling by the tides and which can also be clamped down against the dangers of drowning or drying. Apart from this, many are able to avoid the extremes of their habitat by creeping into crevices, burrowing into sand or hiding beneath stones. Yet in spite of the disadvantages of variability and damaging strength, the sea is not an enemy, for it brings oxygen-charged water to the animals of the shore; to many it acts as a nursery for the young and to some it is the sole supplier of food.

Like land animals, some seashore creatures feed on plants, while others are flesh-eaters. But in the sea there is a third type of food, which is not found on land. This is plankton, which is composed mainly of minute plants and animals, which spend their lives floating in the water. The tides and currents bring a rich haul of plankton to the sedentary animals of the shore, which strain it from the water in a variety of ways, and are referred to as filter-feeders.

Rock pools such as these, with rich growths of algae, teem with wildlife.

The plankton includes some animals which float in the surface of the sea for the whole of their lives, but many are the larval stages of more or less inactive shore creatures, such as sea anemones, crabs or starfishes. These long-lived animals produce vast numbers of minute young, which are carried away on the tides and currents, perhaps to colonise some entirely new area, but certainly not to stay and compete with their parents for living space and food. Their most likely fate is to be eaten at an early stage, but enough survive to ensure continuity of life on the shore.

ROCKY BEACHES

The shores that are richest in animal life are those where rocky headlands break sandy beaches into small coves, for in such a place the differing needs of many sorts of animal will be met. Birds may be the most noticeable inhabitants of such an area. Some, such as the gulls, are land-lubberly scavengers, never leaving sight of the shore and probably moving into the rich safety of towns or fields with the storms of winter. Others, such as fulmars, guillemots, and puffins, which occupy different parts of the cliffs for nesting, are the true seafarers, coming to land only to produce their young, and wintering over the

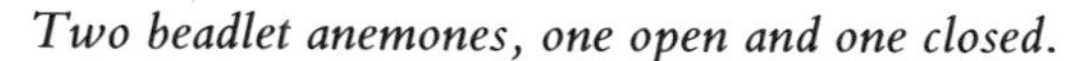

Two beadlet anemones, one open and one closed.

deep oceans. In this they are like the terns which nest on extensive sand or shingle flats, such as those at Scolt Head. In spite of their delicate appearance they are totally at home over the sea and undertake tremendous migrations which ensure that their lives are passed in perpetual summer, for they spend our winter time in the south Atlantic summer.

The permanent inhabitants of the seashore are mostly small invertebrates. On rocks the most obvious are those which are protected by heavy shells, such as barnacles, which may be so closely crowded that it would seem impossible to fit any more in. Several species occur, but they can be difficult to identify, especially the older animals, in which the plates of the shell become worn. At first sight they look like molluscs, but their shelly armour is made up of six strong outer plates and four inner ones which open, at high water, to allow six pairs of feathery legs to be kicked out to comb food from their surroundings. Since they absorb oxygen at the same time, they live very economically, obtaining two essentials by a single activity.

Like barnacles, the limpets are protected by conical shells attached to the rocks. Here the resemblance ends, for a limpet is a mollusc, carrying a single shell and clinging to the substrate with a large muscular foot. This can hold it safely against the most ferocious of

Mussels are often abundant on the seashore and are a major source of food for the oyster-catcher.

Increasingly mussels, as well as oysters, are being raised commercially on artificial reefs since natural mussel beds the size of this one have often been depleted by man.

storms, yet on a summer's day a light tap is sufficient to dislodge it. The reason for this is that the limpet maintains a steady temperature in hot weather by allowing a little of its body fluids to escape and evaporate, for were it to be tightly clamped down it would overheat and die. When the tide rises, limpets move from their home place to feed. They have ribbon-like tongues, closely set with horny teeth with which they scrape small algae from the rocks, but as the tide goes out again, they return to their home spot and orientate themselves so that the edge of their shells and the roughness of the rock match precisely. A small relative of the rock-clinging limpets is the blue-rayed limpet, which can sometimes be found on the stipes of the big brown oarweeds which may be thrown up on the beach after rough weather. This fragile little animal would be unable to stand the harsh conditions of the open beach, but instead scrapes a notch in the seaweed, in which it lies protected. Other sea-snails are less tied to a single abode, although each occupies its own zone of the beach. On the uppermost parts the small periwinkles are air-breathers which feed on shore lichens and breed during the winter months when their eggs can best be carried away by storm-driven high waters. Lower down the beach, the rough periwinkle is succeeded by the brightly coloured flat periwinkle, which live among the bladderwrack and lower still the edible periwinkle may be very common. Other common snails are the top shells. Most of these have a heavy mother-of-pearl layer to the shell which can be seen through the worn outer parts. Like the periwinkles these animals feed on shore plants, but some of the seaside snails are carnivores. They also use a rasp-like tongue to obtain their food, which they devour alive, having in some cases first ground a hole in the shell of their prey. On rocky shores dog whelks feed mainly on barnacles and are usually the most abundant predators, although there are many others, including Britain's two small cowrie shells, which feed on the helpless, soft bodies of colonial sea-squirts of the lower shore.

In rock pools on the middle and lower beach shell-less relatives of the sea-snails may sometimes be seen. Known as sea-slugs, they include some of the most bizarrely beautiful of all creatures, for their soft bodies are often brilliantly coloured and they carry tufts of feathery gills on their backs. In common with many other creatures of shallow seas they come inshore to breed and their spiralling whorls of spawn can often be found in the early summer. The thousands of eggs hatch into planktonic larvae, which for a short time carry a shell, although this is lost when they metamorphose to their adult form. These apparently fragile creatures feed, in most cases, on sea anemones and their relatives and are not only immune to the powerful sting cells of these animals, but can transfer them to their own skins for protection.

Many sea anemones are themselves inhabitants of rocky coastlines. Commonest on the upper shore are the beadlet anemones, so called because of the ring of blue spots which lies at the base of the tentacles. At low tide they look like blobs of green or red jelly stuck on

Left *The snakelocks anemone is found in the lower shore zones since it cannot retract its tentacles and will dry out if exposed by the tide.*

Right *Sea-slugs with their delicate feathery gills are related to the shelled molluscs such as limpets, whelks and winkles.*

The dahlia anemone, photographed here on the side of an aquarium.

the rocks, but as the water rises they expand narrow tentacles armed with poisonous sting cells, for in spite of their name these are animals, and predaceous. Built on a radially symmetrical plan they are ready to receive prey from any direction and paralyse small creatures which brush against their tentacles, and although they have little in the way of sense organs they can distinguish between suitable food and inedible material. The snakelocks anemone sometimes carpets pools on the middle shore. This species gets its name from the fact that it cannot contract its long tentacles, and for this reason also it is always found in water, never in a part of the beach that dries out at low tide. The dahlia anemone which sometimes measures 15cm (6ins) across the expanded tentacles is the largest British species, and one of the most beautiful is the small but vividly coloured *Sagartia elegans*. Related to sea anemones, but much rarer in British waters, are the corals. In pools on the lower shore in south-west Britain the rare Devonshire cup coral or the scarlet and gold star coral may sometimes be found. When expanded they look like small sea anemones, but when contracted the limy skeleton, which distinguishes them from the true anemones, can be seen. Other relatives of the anemones and corals are the hydroids. These are usually small colonial animals, often looking like tiny stalked anemones; while in sheltered places on the lower shore the lobed colonies of polyps called dead man's fingers may occur.

To protect themselves against the fury of the storms and the dangers of desiccation various soft-bodied worms construct limy tubes in which they hide, emerging briefly to snatch food from the water. They are among the wariest of shore animals, alert to the least vibration or shadow of danger, so it is unusual to see them extended. Some animals, however, cannot hide or disguise themselves, but usually live in protected spots, such as the shelter of rocky overhangs, where they are not immediately obvious. They include the sponges, the commonest of which is the breadcrumb sponge. This must be one of the most variable animals in the world, for its colour ranges from that of uncooked pastry through yellows and greens to a pinky red. Nor is the shape constant. On the beach, where it encrusts the surface of rocks, it follows the irregularities of their contours, although in deeper water it may become free-living, developing weirdly contorted shapes. Always, though, the surface is pitted with fine holes through which water is drawn into the labyrinth of the sponge's tissues. Here, individual cells remove oxygen and food particles and spit bodily waste and carbon dioxide into the water, which finally leaves via large apertures set, volcano-like, in the apex of protuberances scattered over the animal's body. Purse sponges are more delicately built and dangle free from overhanging rock. They need the support of the water, for when the tide is out they often collapse to look like small pieces of white rag hung beneath a jut of rock. Other encrusting filter-feeders include the colonial sea-squirts, which in spite of their simple way of life are complex animals, distantly related to the vertebrates. This cannot be seen in the adults, but the larva is an active tadpole-like creature, whose body is supported by a cartilaginous rod, the notochord, which is identical in construction to the precursor of the backbone in developing vertebrates. The golden-star sea-squirt looks like a number of flowers set in a jelly-like substrate, but in reality each apparent petal is an animal which sucks large amounts of sea water through its mouth, though when it is disposed of, minus food and oxygen and plus waste products, it is pumped out of a communal opening in the centre of the 'petals'.

Hiding under stones or beneath curtains of weed, crabs are among the most successful scavengers of the beach. Commonest is the shore crab, whose cast shells are often abundant among the rubbish of the strand line. This is less a testimony to the lack of success of the individuals than an indication of their development, for the crabs and their relatives are all armoured in inflexible but jointed shells, which because they do not stretch to allow growth must be shed periodically. This discarded protective layer gives a complete picture of the crab's surface, for the covering to every hair, the eyes and even the gills is shed. On the middle beach young edible crabs which spend their early life in shallow water can be found, while beneath stones on the lower beach shore the velvet swimming crab lies hidden. It can be recognised by the covering of fine hairs on its back and the elegant navy-blue lines on its legs. When disturbed it comes out fighting, its claws ready to attack, and a nip from them can draw blood. Spider crabs belonging to several genera are found on the lower zones of the beach. Although strongly armoured they are quite inoffensive and make little attempt to defend themselves with their weak claws, but they are disguised with pieces of weed and sponge attached to the shell. Superficially un-crab-like, the common hermit crab is not fully armoured, unlike the majority of shore animals, for the long abdomen is softly fleshy. To protect itself in a cold hard world, the hermit tucks its tender tail into the shell of a dead sea-snail. It fits perfectly, for its body is lop-sided, twisting neatly round the central column of the shell, and bracing itself with small appendages at the rear end. It is impossible to pull a hermit crab from its chosen home without damaging it, though when it needs another shell it vacates the old one easily

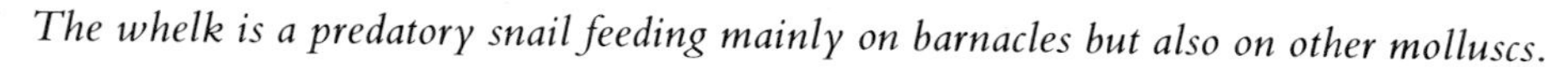

The whelk is a predatory snail feeding mainly on barnacles but also on other molluscs.

enough. It shuts out the world with a large claw folded across the aperture of the shell and is safe from most enemies except the octopus. Hermit crabs sometimes carry sea anemones on their shells and it has been shown that the anemone's sting deters even a hungry octopus. The partnership is an intentional one, for a hermit will stroke the column of a 'parasitic' anemone in such a way as to cause it to relinquish its hold on the substrate, and then place it delicately on its own shell. Both benefit, for the crab gets protection from octopuses and perhaps from other enemies also, and the anemone gets food which drifts into the water as the hermit, an exceedingly untidy feeder, tears at its meals.

On some beaches squat lobsters occur. These heavily armoured creatures have longer tails than crabs. The lobsters and crayfish have yet longer bodies with fan-like tails which enable them to jerk away from enemies at high speed. They live only in the lowest of the rock pools and as they increase in size they retreat to water beyond the range of the tides. The shore pools, however, contain many small creatures which are similar to them, although the relationship is not always as close as their long bodies might suggest at first. The prawns and shrimps are small-scale scavengers. Always alert to the dangers of their world, they drift like ghosts on to any possible food but flick nervously back into the shelter of crevices or weeds at the least untoward movement in the water. Their trans-

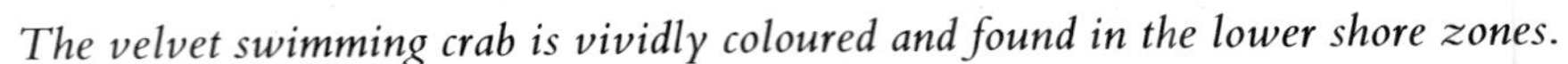
The velvet swimming crab is vividly coloured and found in the lower shore zones.

Left *Fan worms are one of the more attractive species of sandy shores which live much of their lives hidden away in their tubes of cemented sand.*

Right *When covered by the tide limpets, such as these blue-rayed, move around to graze on algae.*

parent bodies make them difficult to see at the best of times, and the habit of burying themselves in the sand makes some of them the most elusive of shore animals. The chameleon prawn changes what little colour it has to match its background more perfectly. Skeleton shrimps hook their thin bodies on to fragments of weed on the lower beach, where they are well camouflaged, while the ghost shrimps (neither is a true shrimp) hang like grey shadows in the water.

On the lowest part of the beach starfishes and their relatives the sea urchins and brittle stars may be found. These animals, collectively known as echinoderms (which means 'hedgehog-skinned'), for most are spiny to some extent, can stand only the briefest exposure to the air, so they never live in the upper zones. Like the sea anemones they are radially symmetrical animals although they have a far more complex structure. The unique feature which unites members of this group is the possession of organs called tube feet. These are like small, water-filled, sausage-shaped balloons, which are protruded through tiny holes in the skin or shell. Most are sucker ended, so they can be used to hold food or to give a good grip as the animals move across their pool. The water is pumped from the sea through a pore on the upper surface of the body and can be withdrawn if the tube foot is no longer needed, so these creatures have great flexibility in the use of their grasping organs. A feature which is apparent in starfishes and brittle stars is their tremendous power of regeneration, for if part of the body is lost it is easily regrown, so many of these animals can be seen with unequal arms. Other members of this group are sometimes found low down on the shore, particularly the sea cucumbers. At first they seem to be worms, but a closer look will show the five rows of tube feet which they carry down the length of their bodies, while the tentacles with which they feed are also modified tube feet.

SANDY BEACHES

While the life of rocky shores is usually easy to see, that of a sandy beach is not, for the sand particles give little for an animal to cling to and no surface shelter for protection. None the less the beach is rich with living things, and an inkling of this may be given by the huge numbers of sandhoppers which sometimes occur on the upper beach where they feed on strand-line detritus. Lower down the shore most creatures remain hidden under the sand, where, apart from the topmost layer, each grain is surrounded by a jacket of water which is the home for countless minute organisms which go largely ignored because of their small size. But the sand also protects many larger creatures which burrow into it. Many kinds of worms, the commonest of which is the lugworm, live in this way and would go unsuspected from the surface were it not for the worm casts and slight depressions in the sand which mark the two ends of the curved burrow.

Even more numerous are the bivalve molluscs which on favourable beaches may number 10,000 to the square metre. They live at different depths beneath the sand, and at high tide push long siphons up to the surface. Sea water is drawn in through these, over gills where the oxygen is removed and the food particles trapped on a band of mucus which transports them to the stomach. Some can, if need be, move across the surface of the sand, but most are helpless if, as sometimes happens, the beach is churned up by a big storm. Besides the weather they have other enemies, some of which they can escape by digging, for many can bury themselves faster than a human inquirer can follow them. Gulls and some wading birds are their most successful predators. The oyster-catcher has a chisel end to its bill with which it chips a hole in the shell of a bivalve; it then severs the muscle holding the shell shut and extracts the soft body. Gulls have a cruder way of dealing with the situation; they dig up shallow-burrowing bivalves such as cockles and carry them to near-by rocks or sometimes, to the consternation of travellers, to roads. Here they drop and smash the shells, swooping down fast to secure their booty, before any of their neighbours get it. Under the sand the unfortunate bivalves have yet other attackers, the worst of which is probably the common necklace shell. This large, rounded snail ploughs through the sand, enfolding with its powerful foot any prey that it encounters. With a strong tongue it rasps through the shell, a process which may take several days but from which there can be only one outcome – a meal for the necklace shell, and later a pair of shells cast upon the beach, neatly bored with a hole about 1mm ($\frac{1}{25}$in) in diameter. Other sand-burrowing snails may be found on the middle and lower beach, including the turret shell, the pelican-foot shell and the beautiful and rare wentletrap, but all of these feed on small organisms. They are seldom seen alive but their shells may be cast up on the shore when they die. Other, larger animals are sometimes found beneath the sand, including the masked crab. This lies just hidden, with its long antennae on the surface making a breathing tube. The heart urchin or sea potato uses its spines to make a deep hole in which it can live securely.

On the surface of the sand shallow pools may be left by the retreating tide and a number of animals can usually be found in them. They include the young of some flat fish which can change their colour to match their surroundings, and large numbers of shrimps and their relatives. Anybody hoping to catch them should take the precaution of wearing shoes, for these pools are the home of the only really dangerous animal of our shores, the weaver fish. This little fish measures only about 12cm (5ins) when it is fully grown, but it is protected against bigger creatures with poison glands connected to spines on the gill covers and the first few dorsal fin supports. Anyone treading on one of these fish will receive an extremely painful wound which could well spoil a holiday.

Occasionally creatures from the open sea are washed up on the shore. They tend to be

Left *Prawns, like many other forms of marine life, are important commercially. However, it is only recently that man has begun to farm the sea and many species have been severely over-exploited.*

Right *Hermit crabs protect their soft bodies by inhabiting the empty shells of dead molluscs, which they change periodically as they grow. Marine biologists have made glass shells for them so that their growth can be observed.*

more noticeable on sandy beaches than on rocky coasts. Some like the comb jelly are so fragile that as they are left by the waves they disintegrate, remaining for a short time as formless blobs of jelly. Others, such as the goose barnacles, can survive for a little while but must have the open sea for food and proper support. Jellyfish are also stranded in this way; the moon jelly, which is commonest is quite harmless, but some of the others can give a very unpleasant sting and are best left alone. So is the Portuguese man o'war, which may arrive on our western coasts after southerly gales have brought it in from the warm oceans which are its normal living place. This should not be touched, for the powerful sting cells will give a painful reminder of these delicate-looking and primitive creatures.

When looking for the animals of the shore it is as well to remember the harshness of their environment and not to add to their problems by heavy disturbance. If stones are turned over in the search for the shyer animals, return them to their original position, for this can mean survival for the animals whose shelter has been disturbed, and by this simple action the richness of life on the shore can be preserved.

Appendix

Nature Reserves and Other Properties of Wildlife Interest Owned by the National Trusts

This list indicates some of the National Trusts' holdings, county by county, which are likely to be of special natural history interest. In fact nearly all the larger estates and stately homes with gardens and parks will have a variety of wildlife present. This guide should be used in conjunction with the various National Trust publications such as *Properties of the National Trust*.

Several of the nature reserves listed here have only limited access, and many are managed by the County Trusts or other specialist bodies. A large proportion have been designated Sites of Special Scientific Interest (SSSI) by the Nature Conservancy Council.

The National Grid reference is given after the name of the property.

ENGLAND

AVON

Dyrham Park (ST 743757)
The National Trust owns the 260-acre park, where fallow deer are kept, 7m N of Bath, 12m E of Bristol, off A46.

Leigh Woods (ST 560734)
160 acres of woodland, which is a National Nature Reserve leased to the Nature Conservancy Council, with many interesting woodland trees and associated wild animals. Near to Bristol, on banks of the river Avon, close to A369.

BEDFORDSHIRE

Dunstable Downs (TL 000190)
Nearly 300 acres, together with 26 acres covenanted by the Zoological Society of London, who own Whipsnade Zoo. In this area Bennett's wallaby, from Tasmania, muntjac and Chinese water deer can be seen. Close to A4146. See also Ashridge, Herts.

BERKSHIRE

Cock Marsh (SU 890869)
132 acres on the south bank of the Thames near Cookham, which is near Maidenhead. (There are several other Trust properties in this area, including Cliveden, Bucks).

Simon's Wood (SU 814637)
63 acres of mixed woodland near Crowthorne.

BUCKINGHAMSHIRE

Boarstall Duck Decoy (SP 624151)
An eighteenth-century duck decoy in working order, between Bicester and Thame, off B4011.

Bradenham Woods (SU 825970)
The Trust owns 1,111 acres, including most of the village of Bradenham, and also some 380 acres of beech and mixed woodlands. 4m NW of High Wycombe, off A4010.

Cliveden (SU 913856)
Parklands and woodlands adjacent to the Thames, totalling 327 acres, with many species of woodland birds, small mammals, and so on. 3m from Maidenhead.

Coombe Hill (SP 849066)
106 acres of downland, rising to 852ft above sea level, the highest point in the Chilterns. 3½m NE of Princes Risborough, south of the B4010.

Waddesdon Manor (SP 740169)
The park contains an eighteenth-century-style aviary and a herd of sika deer. 6m NW of Aylesbury, off A41.

CAMBRIDGESHIRE

Wicken Fen (TL 5570)
The 600 acres are one of the very few remains of the tens of thousands of acres of fen which once spread over East Anglia. An extremely important Nature Reserve off A1123 17m NE of Cambridge. The 125 acres of Chippenham Fen, 3m S of Isleham, have been covenanted to the Trust as a nature reserve (off A142), and are open on application to the Nature Conservancy Council.

Wimpole Hall (TL 336510)
2,443 acres 8m SW of Cambridge. Rare breeds of farm animals.

CHESHIRE

Alderley Edge (SJ 860776)
219 acres of prehistoric mineworkings, beechwoods and a sandstone escarpment. 4½m NW of Macclesfield off A34 and B5087.

Lyme Park (SJ 965845)
1,321 acres of parkland, which includes herds of red deer. SE of Manchester off A6.

Tatton Park (SJ 737815)
The 2,086 acres include a large lake (Tatton Mere) which attracts large numbers of waterfowl. 13m SW of Manchester, off A50.

CORNWALL

With a few exceptions there is a right of way on foot almost the whole length of the Cornish coast, and the National Trust has under its control large sections of the coast-line. For a detailed description see the *National Trust Guide*. The Cornish coastline provides good habitat for breeding sea-birds, while waders are to be found on many of the estuaries and beaches, and seals and dolphins can often be seen offshore. The Trust holdings of coastland and adjacent country-side are too numerous to list here, and the reader should consult the *Guide* and *Properties of the National Trust*.

CUMBRIA

The Trust owns over 122,000 acres and protects nearly 12,000 more.

Blelham Tarn (NY 3600) and adjacent *Blelham Bog*
These are part of the Trust's Hawkshead properties. Their main interest lies in the continuing scientific research carried out. Lord Lonsdale's Commons (NY 3515) comprise nearly 17,000 acres. Near Windermere, off B5286.

Borrowdale Woods (NY 22)
Includes a number of Trust properties, the largest of which is Seatoller Farm (964 acres). Borrowdale probably contains more ancient woodland than any other part of Lakeland. Among the birds can be seen pied flycatcher,

buzzard, wood warbler and grey wagtail. S of Keswick off B5289.

Buttermere (NY 183157)
Although one of the most unproductive waters in the Lake District, fish such as brown trout, charr, salmon, perch, pike and eels are found here. SW of Keswick off B5289.

Other major Trust holdings in Cumbria include Caw Fell, Scafell Pike (the highest point in England) and Langdale Pikes, but for full details of the extensive holdings the reader is referred to the *National Trust Guide* and *Properties of the National Trust*.

DERBYSHIRE

Like Cumbria, large tracts of land are owned or protected by the National Trust, totalling 34,000 acres.

Dovedale (SK 1453)
About 5m NW of Ashbourne, one of the most famous of the Derbyshire dales.

Edale (SK 1383)
Nearly 1,500 acres of upland farms and hills.

Hardwick Park (SK 463638)
The estates of the Hall comprise some 18,000 acres, including 2,300 acres of park adjoining the Hall. The estates and gardens contain a wide variety of woodland wildlife. Off M1, between exits 28 and 29.

Kinder Scout (SK 0889)
The high (636m, 2,070ft) plateau of an upland mire, one of many Trust properties in the Peak District National Park. Good for curlew, golden plover, dunlin and grouse.

DEVON

Arlington Court (SS 611405)
The 2,775-acre estate, of which 200 acres of parkland and woodland contain breeding buzzards and ravens and are visited by the wild red deer of Exmoor and are open to the public. 7m NE of Barnstaple, off A39.

Branscombe and Salcombe Regis (SY 2188, 1588)
Beaches and adjacent coastal cliffs and farmlands. S coast of Devon, E of Sidmouth.

Castle Drogo (SX 721900)
Drewsteignton. Formerly included Whiddon Deer Park but is no longer emparked and contains no deer except wild fallow. Between Okehampton and Newton Abbot, off A382.

Heddon Valley (SS 6549)
941 acres of moorland and coppice oak woodland overlooking the Bristol Channel. E of Lynton, off A39.

Hembury (SX 726684)
347 acres of young oak and coppice woodland in the valley of the river Dart. 2m N of Buckfastleigh, off A384.

Lundy (SS 1345)
1,047-acre island in the Bristol Channel leased to the Landmark Trust. Nearest harbour Ilfracombe. Important reserve with nesting puffins, and other sea-birds, as well as both black and brown rats and a small herd of Soay sheep.

Teign Valley Woods (SX 795885)
Over 200 acres hanging woodland on banks of river Teign. Between Exeter and Moretonhampstead, off B3212.

Willings Walls Warren (SX 6065)
One of several properties which are named after rabbit warrens, once widespread in England. Dartmoor, where they are known as rabbit buries, has good examples of raised artificial warrens: they can be seen on the Trust land at Trowlesworthy. This property is not near any major roads, situated between the A38, A386 and B3212, NE of Plymouth.

DORSET

Brownsea Island (SZ 0288)
In Poole Harbour. 500 acres of heath and woodland with sika deer, red squirrel and other wildlife. Part is a nature reserve leased to the Dorset Naturalists' Trust.

Corfe Castle Estate (SY 959824)
Over 7,000 acres centred on Corfe Castle, the estate contains Purbeck quarries (including caves), heathland, limestone grassland and includes important bird and reptile habitat. 5m NW of Swanage.

Golden Cap Estate (SY 4092)
Nearly 2,000 acres of coast between Charmouth and Eype Mouth, includes a wide variety of habitats. Runs parallel with A35.

Kingston Lacy Estate (SY 978014)
Included in the 8,700-acre estate is the 1,200-acre Holt Heath nature reserve, leased to the Nature Conservancy Council. 2m NW of Wimborne, off B3082.

DURHAM

Moor House Woods (NZ 305460)
Nearly 60 acres of woodland on steep slopes, bordering the river Wear. 3m NE of Durham, close to A1(M).

EAST SUSSEX

Crowlink and Birling Gap (TV 557958)
Coastline west of Beachy Head. 700 acres of chalk cliffs and extensive old grasslands. Off A259, near Friston.

Nap Wood (TQ 585330)
107 acres of oakwood leased to the Sussex Naturalists' Trust as a nature reserve. 4m S of Tunbridge Wells on A267.

ESSEX

Danbury and Lingwood Commons (TL 7805)
Over 200 acres of open common with gorse and birch. 5m E of Chelmsford near A414.

Hatfield Forest (TL 5320)
Over 1,000 acres of one of the oldest forests in England, a former Royal Forest still retaining traces of its origins. 3m E of Bishop's Stortford.

Northey Island (TL 872058)
300-acre island in Blackwater estuary, an important wintering site for birds. Also Ray Island. Near Malden, A414. Leased to the Essex Naturalists' Trust. Access by appointment.

GLOUCESTERSHIRE

Haresfield Beacon and Standish Wood (SO 820089)
354 acres of Cotswold edge, with open limestone grasslands and extensive beechwood. Near A46, 2–3m N of Stroud.

HAMPSHIRE

Hale Purlieu (SU 200180)
500 acres of heath, woodland and valley bog, typical of the New Forest. 3m NE of Fordingbridge, off A338.

Hamble River (SU 523118)
74 acres which include a nature reserve. 1m from Botley off A27.

Selborne (SU 735333)
248 acres of countryside made famous by Gilbert White. 4m S of Alton.

Waggoners' Wells (SU 855350)
A series of old hammer ponds. Off B3002, 1½m W of Hindhead.

Woolton Hill: The Chase (SU 442627)
137 acres of woodland with a chalk stream, managed as a nature reserve. 3m SW of Newbury, off A343.

HEREFORD AND WORCESTER

Brockhampton (SO 682546)
1,680 acres of typical Hereford woodland and farmland. Off A44, 2m E of Bromyard.

Croft Castle (SO 455655)
Parkland, also fish ponds. 5m NW of Leominster, on B4362.

HERTFORDSHIRE

Ashridge Estate (SP 9812)
Nearly 4,000 acres of woods, heaths and

downland. Animals include fallow and muntjac deer and the edible dormouse. 3m N of Berkhamsted, astride B4506.

ISLE OF MAN

Calf of Man
The 600 acres are an important bird reserve with guillemots, chough, seals and other sea-cliff wildlife, leased to the Manx Museum and National Trust.

ISLE OF WIGHT

Newtown Nature Reserve (SZ 4290)
Over 700 acres of estuaries and marshlands and adjoining shores, along with woodland and farmland. Park managed as a nature reserve in conjunction with the Isle of Wight County Council.

St Catherine's Point (SZ 495755)
The most southerly tip of the island, important locality for observing bird migration.

KENT

Appledore Royal Military Canal (TQ 958292)
Marsh frogs and other wildlife in the canal. 8½m NW of New Romney between B2080 and B2067.

Ide Hill (TQ 485515)
32 acres of wooded hillside. 2½m S of Brasted, off B2042.

Knole (TQ 532543)
80 acres. Large park with deer. (Park not owned by the Trust.)

Sandwich Bay (TR 347620)
Nearly 200 acres of saltings and sand dunes with sea-birds and waders. Good area for passage migrants. 2m NE of Sandwich, off A256.

Scotney Castle Garden (TQ 688353)
Nearly 800 acres, which include a wide variety of gardens and habitats.

LANCASHIRE

Eaves Wood (SD 465758)
100 acres of wooded hill, E of Castlebarrow Head, near A6. Public car park.

LEICESTERSHIRE

Charnwood Forest (SK 490126)
Ulverscroft Nature Reserve managed by Leicester and Rutland Trust for Nature Conservation. Permits required away from public footpath. 6m SW of Loughborough, between B5350, B591 and B587.

LINCOLNSHIRE

Belton Park (SK 932386)
700 acres of parkland, 3m N of Grantham.

LONDON

East Sheen Common (TQ 197746)
Adjacent to Richmond Park, close to S Circular Road.

Petts Wood (TQ 450687)
88 acres of woodland and heath with good range of woodland birds. Near Bromley, off A208.

Selsdon Wood (TQ 357615)
Nearly 200 acres of woodland on outskirts of suburbs with badgers, and so on. 3m SE of Croydon, near A2022.

MERSEYSIDE

Formby (SD 275080)
Over 400 acres of sand dunes and pine woods. Near Liverpool, off A565, at Formby.

NORFOLK

North Norfolk contains some of the Trust's most important nature reserves. The reader is referred to the National Trust booklet on the North Norfolk Coast for more detailed information.

Blakeney Point (TG 0046)
1,100 acres of shingle spit, saltmarsh and sand dunes; an important nature reserve 8m E of Wells, off A149.

Blickling (TG 1728)
Over 4,700 acres of park, farm and woodland. 1m NW of Aylsham, on the A140.

Brancaster and Scolt Head (TF 800466)
Over 3,000 acres of beach, foreshore, dunes and saltmarsh. Interesting marine life and tern colonies. Between Hunstanton and Wells.

Stiffkey Marshes (TG 9543)
487 acres of marsh stretching along 2 miles of coast. Between Blakeney and Wells, off A149.

Other near-by Trust-controlled lands include Morston Marshes and Salthouse Broad (Arnold's Marsh).

NORTHAMPTONSHIRE

Ashton Wold (TL 088878)
Over 500 acres covenanted. 3m E of Oundle.

NORTHUMBERLAND

Allen Banks (NY 799630)
3m W of Haydon Bridge S of A69. Nearly 200 acres of wooded riverside banks and crags. 3m W of Haydon Bridge S of A69.

Cragside, Rothbury (NU 073022)
Over 2,300 acres, including 900 acres of wooded park. A country park, 16m NW of Morpeth, entrance from B6334.

Farne Islands (NU 2337)
30 islands covering 80 acres. One of the Trust's most important nature reserves, famous for its seals and sea-birds. Reached by boat from Seahouses. Open to visitors during the summer only.

Newton Pool (NU 243240)
16½ acres near Newton-by-the-Sea, adjoining Embleton Links. Freshwater pool, nature reserve, particularly for birds. Access to edge of reserve by footpath from Newton-by-the-Sea or Embleton. Hide for disabled visitors with path for wheelchairs.

NORTH YORKSHIRE

Bridestones Moor (SE 8791)
280 acres of nature reserve. 12m S of Whitby, 1m E of A169.

Hayburn Wyke (NZ 010970)
65 acres of nature reserve, 6m N of Scarborough on coast.

Malham Tarn Estate (SD 8966)
Over 4,000 acres of a wide variety of habitats, part leased to the Field Studies Council. 6m NE of Settle (A65).

Ravenscar (NZ 980025)
Over 250 acres along 1 mile of coastline.

NOTTINGHAMSHIRE

Clumber Park (SK 6477)
Over 3,700 acres including lakes, park and woodland. A country park 2½m SE of Worksop. Excellent area for birds, including waterfowl on large lakes.

OXFORDSHIRE

Ruskin Reserve (SU 459997)
4½ acres of marshy woodland leased to the Nature Conservancy Council. 3m NW of Abingdon.

Watlington Hill (SU 702935)
96 acres of open down and copse on an escarpment of the Chilterns. 1m SE of Watlington.

SHROPSHIRE

Attingham Park (SJ 542093)
Over 3,800 acres including a deer park with fallow deer. 4m SE of Shrewsbury on N side of A5.

SOMERSET

Cheddar Cliffs (ST 468543)
375 acres including 100 acres of Gorge,

important for bats and with good range of butterflies. 8m NW of Wells.

Ebbor Gorge (ST 525485)
142 acres; a nature reserve with caves, leased to the Nature Conservancy Council. 3m NW of Wells.

Exmoor (SS 8844)
The Trust controls nearly 14,000 acres, including the open moorland round Dunkery Beacon and the Horner Valley oakwoods. Wide range of habitat supports a rich and varied fauna, including red deer.

SOUTH YORKSHIRE/DERBYSHIRE

Derwent Estate (SK 1994)
Nearly 6,500 acres in the Peak District National Park. High moorland with good red grouse and other moorland birds.

STAFFORDSHIRE

Hawksmoor (SK 0344)
307 acres of undulating woodlands and open country: 1½m W of Cheadle.

Manifold and Hamps Valley (SK 0153)
Over 300 acres of woodland, open limestone grassland and limestone cliff in valley of Manifold and Hamps rivers. About 7m NW of Ashbourne.

SUFFOLK

Dunwich Heath (TM 475683)
Over 200 acres close to the important reserves of Minsmere and Walberswick. Close to Dunwich and Westleton, near A12.

SURREY

Bookham and Banks Common (TQ 1256)
Nearly 450 acres of woodland and common, which have been studied in detail for over 40 years. 2½m W of Leatherhead, adjacent to Bookham Station, and between A245 and A246.

Box Hill (TQ 1751)
Nearly 1,000 acres of down and woodland. Field Studies Centre at Juniper Hall, leased to the Field Studies Council. Also 100 acres of Mickleham Down and 150 acres of Ranmore Common. 2½m S of Leatherhead, E of A24.

Frensham Common (SU 8540)
Nearly 1,000 acres of heathland and ponds. Once breeding site for natterjack toad. Astride A287.

Hindhead
Several woodland and heathland properties are owned in the area around Hindhead and Haslemere, and in adjacent Hampshire.

Leith Hill (TQ 139432)
The highest point in SE England (965ft), several acres of heath and woodland owned by the Trust. Near to A29, A24 and A25.

Limpsfield Common (TQ 410525)
350 acres of common, heath and woodland SE of Oxted on border with Kent.

Reigate (TQ 250520)
150 acres of open downland and beech woods W of A217.

Witley and Milford Commons (SU 9240)
240 acres of common, with Information Centre for countryside interpretation. Between A3 and A286, ½m SW of Milford.

WARWICKSHIRE

Charlecote Park (SP 263564)
Over 200 acres of parkland with fallow and red deer alleged to have been poached by Shakespeare, also Jacob's sheep introduced in 1759. 4m E of Stratford upon Avon.

WEST SUSSEX

Black Down (SU 9230)
Over 600 acres including relict heathland. 1m SE of Haslemere.

Petworth (SU 976218)
The herd of fallow deer painted by Turner in the grounds is now owned by the Trust. 5½m E of Midhurst.

West Wittering: East Head (SU 766990)
76 acres of dunes and saltings, part of the more extensive Chichester Harbour estuarine habitats. Waders and sea-birds. E side of Chichester Harbour.

WEST YORKSHIRE

Hardcastle Crags (SD 9930)
The valley woods (400 acres) round Hardcastle Crags are managed by Calderdale District Council. 5m NE of Todmorden.

Marsden Moor (SE 0210)
Over 5,600 acres of moorland, southern part in Peak District National Park.

WILTSHIRE

Stourhead (ST 7735)
Over 2,600 acres of parkland and gardens, with lakes, woods and downland. Stourton, on B3092.

NORTHERN IRELAND

COUNTY ANTRIM

Carrick-a-Rede and Sheep Island (D 062450)
Salmon fishery; Sheep Island important for nesting birds. 5m W of Ballycastle and 55m from Belfast.

North Antrim Cliff Path
10-mile cliff-top walk from Giant's Causeway to Whitepark Bay.

COUNTY DOWN

Blockhouse and Green Islands (J 2509, 2411)
Two islands near Cranfield managed as reserve by the Royal Society for the Protection of Birds. Nesting colonies of terns.

Cockle Islands (J 5283)
½-acre island near Ballymacormick; no access during breeding season.

Murlough National Nature Reserve (J 410350)
Over 900 acres, including sand dunes and beach, common seals. 2m NW of Newcastle and 28m S of Belfast.

Strangford Lough Wildlife Scheme (J 500615)
An extensive scheme embracing all the foreshore and islands of Strangford Lough. Viewpoints, hides and walks are provided. Important wintering wildfowl with brent goose, nesting colonies of terns, seal breeding and generally good for waders.

COUNTY LONDONDERRY

Bar Mouth (C 792355)
19 acres of nature reserve, with observation hides.

WALES

DYFED

Dolaucothi (SN 6640)
Over 2,500 acres, which include old gold-mines with bats; fallow deer may occur. Between Llanwrda and Lampeter.

Marloes (SM 758091)
Over 70 acres including the old Deer Park, Martin's Haven and Gateholm Islands. St Brides Bay.

Stackpole (SR 977963)
Nearly 2,000 acres which includes lakes, woods and 8 miles of cliffs and dunes. Wide habitat range, with sea-birds, good woodland and freshwater lake interests. 4m S of Pembroke.

GWENT

Sugar Loaf (SO 2718)
Over 2,000 acres, mainly common. 1½m NW of Abergavenny.

GWYNEDD

Aberglaslyn Pass (SH 600468)
Over 500 acres which includes old copper mines (dangerous). Astride A487 and A498.

Carneddau (SH 6760)
15,800 acres of Snowdonia. 8m SE of Bangor.

Cemlyn (Anglesey) (SH 325933)
2m of coast. Nature reserve. 2m W of Cemaes Bay.

Coedydd Maentwrog (SH 670418)
Over 200 acres of oak wooded valley. Nature reserve leased to the Nature Conservancy Council. 7½m E of Porthmadog.

Ysbyty Ifan (SH 8448)
25,800 acres of Snowdonia, with Llyn Conwy. Moorland birds including red grouse. S of Betws-y-Coed.

POWYS

Brecon Beacons (SO 0120)
The Trust owns over 8,000 acres including Penyfan.

Henrhyd Falls and Graigllech Woods (SN 850119)
37 acres of wooded ravine 11m N of Neath.

WEST GLAMORGAN

Llanrhidian Marsh (SS 490932)
1,200 acres of saltmarsh, important for wildfowl and waders with extensive feeding areas adjacent. 6m W of Swansea, via B4295.

Worms Head (SS 383878)
1,000 acres of the headland. Managed by the Nature Conservancy Council as a nature reserve.

SCOTLAND

The National Trust for Scotland is a separate organisation.

ARGYLLSHIRE

The Trust owns nearly 13,000 acres of rugged highland country in Glencoe and Dalness.

AYRSHIRE

Culzean Country Park (NS 230103)
560 acres of Scotland's most popular Trust property. A wide diversity of habitats and wildlife, including shorelife.

BERWICKSHIRE

St Abb's Head (NT 913674)
An important breeding site for razorbills, guillemots, shags, kittiwakes and fulmars. 192 acres.

DUMFRIESS-SHIRE

Grey Mare's Tail (NT 186146)
A waterfall and surrounds, between Loch Skeen and Moffat Water, where wild goats are occasionally seen.

FAIR ISLE; SHETLAND

An important bird observatory, and breeding grounds for skuas and other sea-birds. 60m NNE of Orkney.

HIGHLAND

Canna and Sanday
The most westerly of the Small Isles. 3,700 acres, including interesting birdlife. Access by boat from Mallaig.

ISLE OF ARRAN

6,600 acres of mountainous countryside with one of Britain's largest lists of breeding birds of prey, including golden eagles, merlin and peregrines; also red deer.

ISLE OF MULL

Burg
1,500 acres of a remote peninsula, with feral goats and cliff-nesting birds.

KIRKCUDBRIGHTSHIRE

Rockcliffe (NX 847537)
100 acres close to islands with nesting sea-birds (closed to visitors during breeding season). Coastal birds and shorelife.

PERTHSHIRE

Ben Lawers (NN 608379)
7,500-acre estate with 1½hr nature trail. Botanically important but also much other wildlife.

Linn of Tummel (NN 914610)
50 acres with a forest nature trail (2m); birds include dipper, grey wagtail and redbreasted merganser by the river.

ROSS AND CROMARTY

Torridon
A total of 16,000 acres adjacent to Beinn Eighe. One of the last strongholds for much of Britain's wildlife including pine marten, red squirrel, wild cat, golden eagle, red deer, otter and seals.

WESTERN ISLES

The St Kilda Group is one of Britain's most important nature reserves, particularly famous for its sea-birds and primitive Soay sheep. There are also local races of mice and wren. 60m W of Harris.

WESTER ROSS

The Trust has large holdings, viz Kintail (12,800 acres), Balmacara (5,600 acres) and Falls of Glomach (2,200 acres) which preserve an impressive section of highland scenery and wildlife including golden eagles, red deer and wild goats.

1 South-west England and Wales

2 South-east and central England
N
West Yorkshire
Marsden Moor
South Yorkshire
Kinder Scout
Derwent Estate
Sheffield
Edale
Clumber Park
Derbyshire
Lincoln
Hardwick Park
Lincolnshire
Manifold and Hamps Valley
Nottinghamshire
Dovedale
Nottingham
Belton Park
Boston
Brancaster and Scolt Head
Blakeney Point
Stiffkey Marshes
Wells
Blickling
Hawksmoor
Derby
King's Lynn
Norfolk
Staffordshire
Charnwood Forest
Norwich
Leicester
Peterborough
Lowestoft
Leicestershire
Ashton Wold
West Midlands
Thetford
Cambridgeshire
Dunwich Heath
Wicken Fen
Suffolk
Northamptonshire
Warwickshire
Northampton
Cambridge
Bury St Edmunds
Charlecote Park
Wimpole Hall
Bedfordshire
Ipswich
Buckinghamshire
Bedford
Harwich
Gloucestershire
Hertfordshire
Colchester
Waddesdon Manor
Hatfield Forest
Dunstable Downs
Oxfordshire
Danbury and Lingwood Commons
Cheltenham
Oxford
Boarstall Duck Decoy
Ashridge Estate
Northey Island
Coombe Hill
Chelmsford
Essex
Ruskin Reserve
Bradenham Woods
St Albans
Watlington Hill
Cock Marsh
Cliveden
Berkshire
Maidenhead
Greater London
Margate
Simon's Wood
Petts Wood
East Sheen Common
Wiltshire
Woolton Hill: The Chase
Selsdon Wood
Canterbury
Sandwich Bay
Bookham and Banks Common
Ide Hill
Knole
Maidstone
Box Hill
Witley and Milford Commons
Guildford
Reigate
Limpsfield Common
Kent
Hampshire
Frensham Common
Surrey
Scotney Castle Garden
Dover
Leith Hill
Hindhead
Nap Wood
Waggoners' Wells
Appledore Royal Military Canal
Selborne
Black Down
Petworth
Southampton
East Sussex
Hamble River
West Sussex
Hastings
Portsmouth
Chichester
Brighton
Eastbourne
Newton Nature Reserve
East Head
Crowlink and Birling Gap
Isle of Wight
St Catherine's Point

3 The north of England

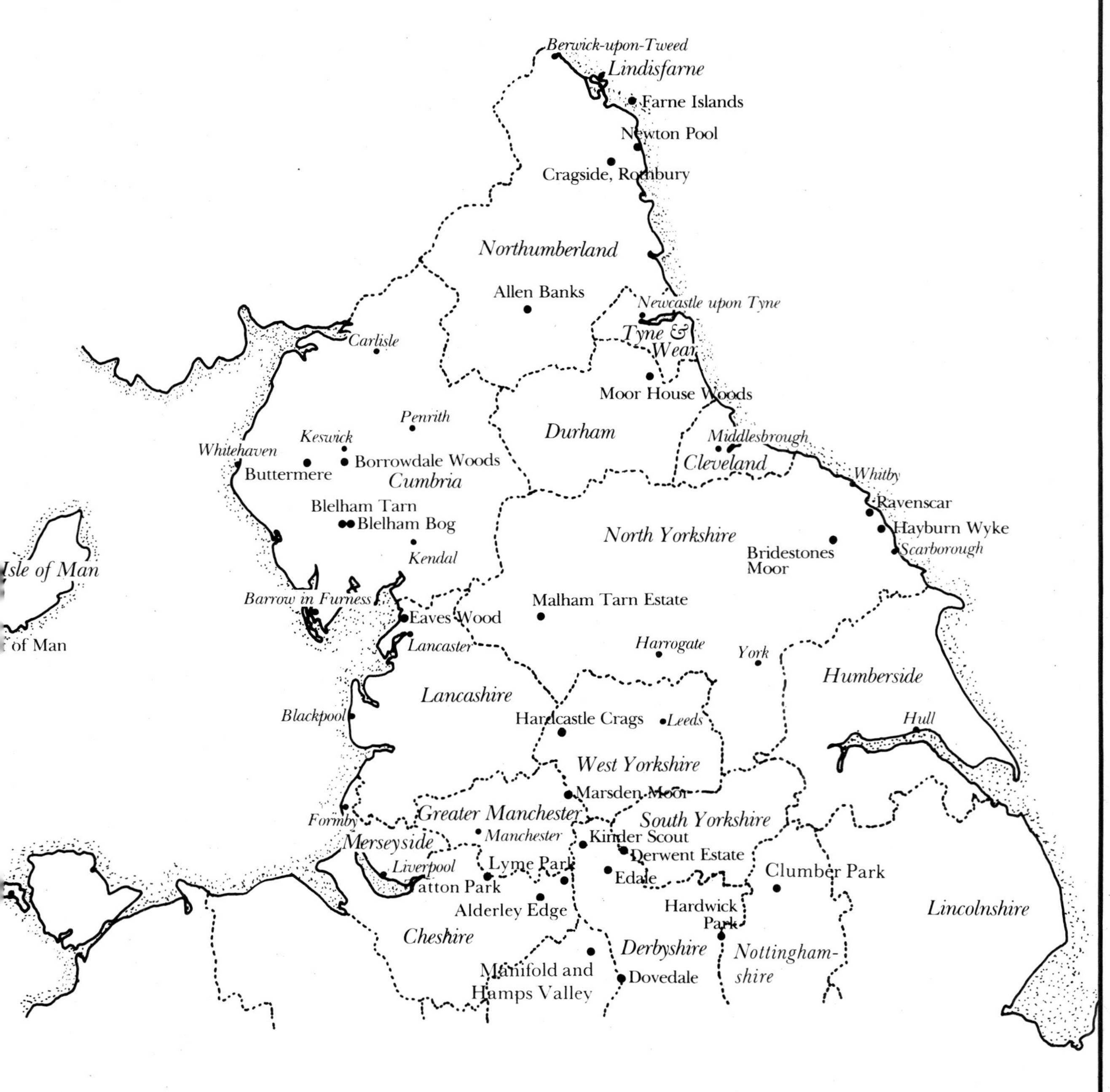

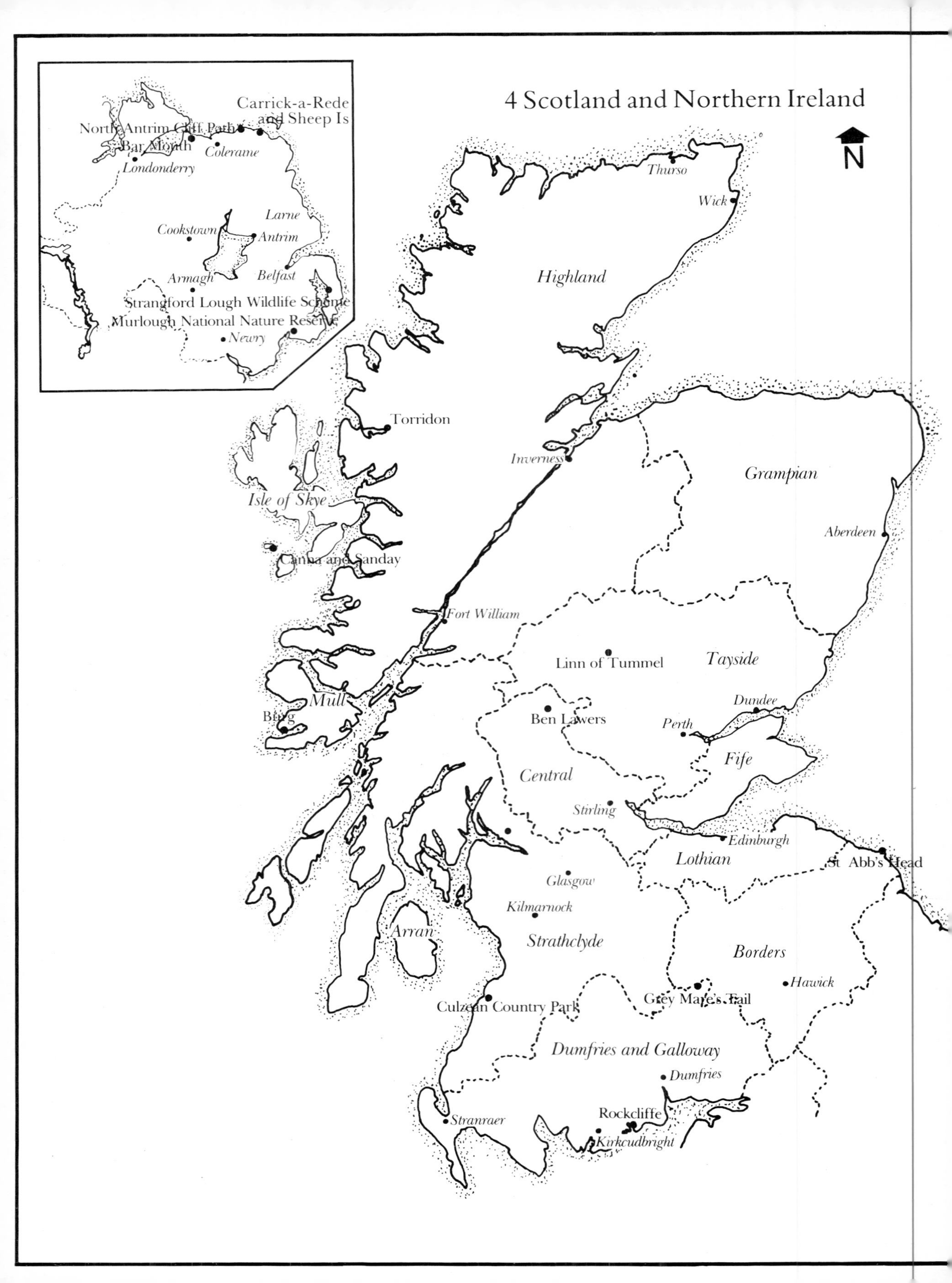
4 Scotland and Northern Ireland
N
Carrick-a-Rede and Sheep Is
North Antrim Cliff Path
Bar Mouth
Coleraine
Londonderry
Larne
Cookstown
Antrim
Armagh
Belfast
Strangford Lough Wildlife Scheme
Murlough National Nature Reserve
Newry
Thurso
Wick
Highland
Torridon
Inverness
Grampian
Isle of Skye
Aberdeen
Canna and Sanday
Fort William
Linn of Tummel
Tayside
Dundee
Mull
Burg
Ben Lawers
Perth
Fife
Central
Stirling
Edinburgh
Lothian
St Abb's Head
Glasgow
Kilmarnock
Arran
Strathclyde
Borders
Hawick
Culzean Country Park
Grey Mare's Tail
Dumfries and Galloway
Dumfries
Rockcliffe
Stranraer
Kirkcudbright

Further Reading

FIELD GUIDES

There are various field guides to most groups of animals found in Britain. Among the most comprehensive and best are the following:

Arnold, E.N., Burton, J.A., and Ovenden, D., *The Amphibians and Reptiles of Britain and Europe* (Collins, 1978).

Barrett, J.H., and Yonge, E.M., *Collins Pocket Guide to the Sea Shore* (Collins, 1958).

Chinery, M., *Field Guide to the Insects of Britain and Northern Europe* (Collins, 1973).

Corbet, G., and Ovenden, D., *The Mammals of Britain and Europe* (Collins, 1980).

Heinzel, H., Fitter, R., and Parslow, J., *The Birds of Britain and Europe* (Collins, 1972).

Higgins, L.G., and Riley, N.D., *Field Guide to the Butterflies of Britain and Europe* (Collins, 1980).

Kerney, M., and Cameron, R.A.D., *Field Guide to the Snails of Britain and North-West Europe* (Collins, 1979).

Muns, B., *Freshwater Fish of Britain and Europe*, ed. Alwyne Wheeler (Collins, 1971).

For insects, the 'Wayside and Woodland Series' published by Frederick Warne, though mostly only available secondhand, is excellent.

HANDBOOKS

Cramp, S. (ed.), *Handbook of the Birds of Europe, the Middle East and North Africa* (Oxford, 1977–).

Corbet, G.B., and Southen, H.N., *The Handbook of British Mammals* (Blackwell Scientific Publications, 1977).

GENERAL

The 'New Naturalist' series, published by Collins, is by far the most authoritative series on the wildlife and habitats of the British Isles.

OTHER NATIONAL TRUST PUBLICATIONS

The National Trust Atlas, 2nd edn (National Trust and George Philip, 1984).

Fedden, Robin, and Joekes, Rosemary (ed.), *The National Trust Guide*, 3rd edn (Jonathan Cape, 1984).

Prentice, Robin, *The National Trust for Scotland Guide*, rev. edn (Jonathan Cape, 1981).

Index